HOW TO FEEL

HOW TO FEEL

THE SCIENCE AND MEANING OF TOUCH

SUSHMA SUBRAMANIAN

HarperCollins *Publishers* India

First published in New York in 2021 by
Columbia University Press

First published in India by
HarperCollins *Publishers* 2021
A-75, Sector 57, Noida, Uttar Pradesh 201301, India
www.harpercollins.co.in

2 4 6 8 10 9 7 5 3 1

P-ISBN: 978-93-5422-700-4
E-ISBN: 978-93-5422-602-1

Printed and bound at
Replika Press Pvt. Ltd.

MIX
Paper from responsible sources
FSC® C016779
FSC
www.fsc.org

This book is produced from independently certified FSC® paper to ensure responsible forest management.

Why was the sight

To such a tender ball as the eye confined

So obvious and so easy to be quench'd?

And not, as feeling, through all parts diffused,

That she might look at will through every pore.

—John Milton, from *Samson Agonistes*

Contents

Introduction *ix*

1 Dull: How Our Cultures Lost Touch *1*

2 Numb: Life Without Touch *25*

3 Mushy: When Sensation Crosses Into Emotion *49*

4 Untethered: Will the Body Become Obsolete? *73*

5 Softening: Overcoming Touch Aversion *96*

6 Boundaries: Knowing Good Touch from Bad *119*

7 Slick: How Companies Sell Us Touch *143*

8 Haptics: Bringing Touch to Our Technology *167*

9 Tactful: Building Machines That Can Touch Us Back *190*

Conclusion *208*

Notes *221*

Bibliography *233*

Index *245*

About the Author *249*

Introduction

There were a long few years when I was making a living as a full-time magazine freelancer that I was mostly without significant contact with another person. I worked alone. I ate alone. I went to bed alone. I had roommates who were busy with their jobs and with whom I'd chat on their way in and out, and I had a circle of people I'd meet with in the evenings, but most of my dearest friends and family were a phone call away. I spent the majority of my time in my head and in my work, which isn't to say I was lonely or depressed. I enjoyed it. I love spending time by myself. But I started to wonder if it even mattered that I was there, that I existed.

I asked myself why I even lived where I did, the way I did. I could have moved anywhere and continued in pretty much the same fashion. My daily practices reflected the way I thought about my physical presence in the world. Like many people who work from home, I lived in sweats. It didn't matter if I showered. It became easy for me to forget to drink water or to sleep since I was so absorbed in the screen in front of me. I had to set reminders on my phone to get up and exercise so I could remain a functioning human. I did it because I knew it was good for me, but I hardly ever felt the urge. There was no inner voice telling me to go out and remind myself my body was alive.

This pattern of daily life was my norm when I started dating my now-husband, Kartik. I met him the way most people do these days, which is online. We're both inherently reserved people, so while we could charm our way through our first conversations, physical contact came slowly. We inched closer and closer to each other while sitting on a couch. I'd place a hand on his shoulder when he told a joke. I remember when we held hands for the first time while together at his apartment. It was like an electrical current ran up my arm and dislodged something in my throat. I didn't know how much I needed that touch, how hungry my skin was for it. Something in me came awake that day. I ate better. I slept better. I've gone back to that moment repeatedly while writing this book, and now I see how much of what I was feeling is so common in today's environment.

How we use our senses isn't just biological. It's shaped by lessons from our culture. In the West, we favor vision, which means that physical presence, which touch epitomizes, fades away. While we've surrounded ourselves with interesting sights and sounds, our tactile existence has become comparatively bare. We don't very often feel the satisfaction of building something with our own hands or the revolving temperatures of the seasons or the security of a doctor consoling us when we're sick. Since we live without these sensations, we hardly recognize that we miss them, and we don't think about what it would mean to try to preserve them. To understand how we got here, we have to go back hundreds of years, all the way to the ancient Greeks.

Ever since Plato, touch has been an unappreciated sense. For him, ranking the senses from highest to lowest was a central exercise in examining what makes humans different from the rest of the animal kingdom. If an aptitude for higher thought is what sets us apart from the lower creatures, which have only basic bodily awareness, then the senses could be hierarchically arranged, from those that are the most aligned with the mind to those most concerned with the corporeal. Vision, as a way of experiencing the world that operates at a physical remove, allows for cool reason, Plato thought. He put it at the top. Touch is immediate and visceral. It is concerned with gut-level needs, basic survival, and sexual temptation. It ranked as the lowest.

But even as Plato denigrated touch, his feelings about it weren't absolute. He used it to describe the moment when the soul made contact with the divine. Aristotle thought the universality of touch was an indication of its importance and power; it's how animals are able to recognize their own existence, and humans' superior thought helped to form a capacity for mental self-reflection. Touch has been considered by philosophers, writers, and artists as a powerful sense because it tells us the truth when our eyes deceive us. If vision is associated with knowledge, then touch is most representative of affect. When we meet an unfriendly person, we describe them as "cold." When an experience moves us, it is "touching." We draw these links in part because our emotions are expressed through physical changes in our skin. If we subtract all the other bodily sensations from an emotion such as anger—the pulsing of a heartbeat, the deepening of the breath, the heat rising in our skin—it's doubtful whether the emotion would even be recognizable to us anymore.

We've carried these mixed messages about touch with us through the ages, and we perpetuate them without realizing it. Plato's mind/body dualism fed into Christian moralizing of vision as beautiful and godly and touch as dirty, the seat of our passions. At the same time, there are numerous examples of divine touch for healing. During the Enlightenment, methods of learning that used vision, such as reading and observing, were idealized. The act of seeing became symbolic of knowledge itself. Concurrently, as touching acquaintances became rarer, literature romanticized subtle touches between love interests. They were portrayed as magical and soul-churning. The Industrial Age was a period that saw the replacement of hand-held tools with large machines, and it launched a counter-response—the Arts and Crafts movement, which aimed to bring back a respect for the intricacy and imperfection of human handiwork.

These contrasting feelings about our sense of touch, as an unsophisticated sense but also one with unique access to the psyche, molds our products and practices today. Our machines are mostly flat boxes with pictures, and there are few moments when we aren't gazing upon their blue glow. They don't ask us to push levers and buttons and materially interact with them the way our machines did even thirty years ago. These same machines are sold to us as a way to stay constantly connected to our loved ones and to build community. Tactile words are even used to make it seem more human: iPod Touch and HP Touch Pad. We form our egos around our highly curated visual personas that we exchange on them, but these images don't compare to being together in the flesh. We're conflicted about what our computers and cell phones do for us—keep us together or strip us of intimacy and empathy.

Our ambivalence regarding touch is observable in the way we conduct our relationships. Smaller households, delayed marriage, and longer life expectancy mean we don't have the physical closeness that we would have expected in the past. As touch has become sexualized in American culture in the past few decades, we've demonized it in most other personal and professional settings. Fears about lawsuits and the rise of the #metoo movement have made us avoid any interactions that might hint toward inappropriateness. Because affectionate touch tends to be reserved for only for our most intimate relationships—with exceptions depending on our culture, gender, and personality—many people don't receive any. It's not yet clear whether our etiquette is protecting us or alienating us from each other.

Several prominent voices are crying out about how our growing detachment is making us less able to handle the challenges of living and working beside other people. We openly talk about our epidemic of loneliness, including our lack of tactile bonding, and how it's leading to chronic illness. Our era has even been termed as the Age of Excarnation, a time when we're living outside of our bodies, meaning we observe ourselves more as externalized images than as living skin and bone. When we put a finger on the pulse of the culture, we find that our daily lives are telling us that we're disembodied eyes and brains. Virtual reality beckons as a final invitation to leave our lived environments and escape into another without any of the friction of real life.

We have a deep anxiety about an increasingly visual world. While we've long been reckoning with these issues personally and socially for centuries, today a vast area of research is reminding us how important touch is. When we're babies, our parents' touch is how we know that we're cared for, and it's crucial to our physical growth and emotional development. Through it, we develop our understanding of what it means to be separate and yet connected. Touch is our first way of exploring the world. As we bump up against our surroundings, we get a sense of the limits of our bodies and how we can use them to extend our will.

The subtle signals from touch help us learn to walk, to pick up objects, and to solve our first basic problems. The senses of vision and touch act as a check-and-balance system through which we arrive at a compromise between our peripheral realities and our deeper wants.

The sensations we receive through touch form a profound form of knowledge that usually take place outside our consciousness, which is why we fail to appreciate them. Many scientists believe that working with our bodies, particularly our highly sensitive hands, is how humans laid the groundwork to build technology and the framework for symbolic thought, and perhaps even our language. In other words, we built atop our skills for physical invention the ability to tinker and toy with ideas solely in our minds. Some believe we're doing ourselves a disservice by failing to use the hands-on learning that served us so well in the past. Children especially might lack engagement when they're so focused on reading and memorizing books.

Caring, friendly touch helps us act more cooperatively and curbs our more aggressive tendencies. In studying the growing epidemic of loneliness in Western countries, psychologists increasingly focus on our lack of touch and not just the problem of having no one to talk to. We're recognizing that some people have such a deep need for physical contact that a lack of it can contribute to depression, trouble empathizing with others, and a compromised immune system. This collection of symptoms is seen in a condition that has been called skin hunger. Avoiding touching altogether may make some social situations easier, but if we don't allow ourselves some of the discomfort of closeness, then we also can't experience its rewards. We forget the importance of small pats on the back and hand squeezes. It isn't until we go without them for a long time and someone touches us again that we even notice what we're missing.

Even small touches can elicit meaningful emotions and affect our values, and consumer goods companies are taking notice. They are using the latest research on how the sense works and how to appeal to it, incorporating new materials and designs meant to increase our enjoyment of our products. The quirky field of haptics, which is trying to bring more physicality to our technology, is now experiencing a golden age that could change the way we all use our sense of touch with our devices. Haptics engineers are thinking about how to replace the feelings, such as typing and the turning of dials, that we used to get from our machines so we have a deeper connection to them. The popularity of touch screens, which allow us to interact directly with images, offers up possibilities for an entirely new library of sensations to develop.

I don't remember exactly when I started thinking about touch. But one vague memory sticks out, when I was procrastinating on my work one day. The top of my desk was feeling a little loose, so I stood up and tried to lift it off its base to see which one of the screws needed tightening. As my fingers curled around the corner, I noticed all the information that I was taking in, from the graininess of the wood, to the strain of my muscles, to the pinching of my skin. All of these subtle cues helped me figure out the answer. I started to wonder which of these sensations were part of my sense of touch. Was it just what I felt on my skin? Or was it all of them together?

It was unsettling to realize how little I knew about my sense of touch, considering it's a capacity that never shuts off. What I'd learned back in school felt profoundly incomplete. Since kindergarten, I'd thought of it simply as how I *feel*. It's how I could tell without looking that I'd landed my feet securely on the pedals of my bike. It was the itch of a bug bite. It was the agonizing pain of stubbing my little toe. It was the scratchy kiss of an overworn sweater. It was the relief of leaning my head on a friend's shoulder. But coming up with a working definition of touch is far more complicated than it might seem initially. In fact, even scientists still disagree on how to go about it. They're not even sure if it's just one sense or a cluster of many.

Although touch has always been a sense with a shaky definition, these debates became much more prominent with the invention of electron microscopy in the twentieth century, which has made it possible to see the wide array of structures in our bodies that are devoted to our sense of touch. Of all the senses, touch continues to be the least understood because it has so many components that are difficult to isolate. But it's not just science that gets to characterize touch for us. Classical Western philosophy has been committed to the idea of five basic senses for centuries. According to our cultural understanding of touch, it seems obvious to us that temperature and itch and pain and pressure are all parts of touch and help create our perceptions of our bodies' movement through space. This is how we'll examine it here.

Touch was an unexpected subject for me, mostly because I've never really liked it that much. When I was younger, my father used to call me a "touch-me-not" after the fern that folds in on itself when it's stroked because I had a tendency to shrink away when anyone came too close. I've never developed the ease with my body that some people have with theirs, although I've since mostly outgrown my extreme aversion. I'm happy to shake hands or hug if someone else initiates. There are some people I do touch voluntarily. I like to touch Kartik. But I think my lifelong squeamishness around touch is another reason I was drawn to this subject. It probably takes someone with strong feelings around touch to even notice it since it's a sense that so often operates in the background.

Growing up as a child of Indian immigrants, I was also acutely aware of the way my tactile interactions with my family differ from those of my peers and their families. We ate rice with our hands instead of a fork and nodded our heads to greet relatives instead of hugging them. I got the impression that how my two cultures use touch differently says a lot about them. When I switched between these two ways of acting, I noticed small shifts in my thinking about issues such as hygiene and politeness and how to exist as an individual while being part of a group. The way we use our bodies is so often a physical representation of our internalized states of being.

I started this project with basic questions about what touch is and how it works. Over time my questions became more personal and more pressing. I began considering all of the cultural programming that led me to my unique relationship with touch, numerous examples of which are fairly universal. When we're young and are warned to "keep your hands to yourself" or told that we are "just being sensitive," we're conditioned to think of touch—and its outgrowth, our feelings—as crass and weak. By examining these deep biases, I redefined touch for myself as an important and wise sense. I appreciated how getting a massage grounded me in my body and how writing with pen on paper helped me work out problems I couldn't solve in my head. I started considering what our world would look like if we all examined our complicated relationship with touch and could relearn to embrace it.

In my research for this book, I've looked at how our culture at large has changed. I recognize there are problems with using the plural "I" to represent the collective "we" to describe behaviors in the United States or the West. The intention is not to leave anyone out or suggest that there's no variation between us; it's to highlight how our individual practices come together on a broader scale. There is, of course, significant variation in the touch practices of individuals depending on their personality, social class, level of health, gender, sexuality, and general background. So in several chapters I explore many types of bodies, including those that are marginalized.

An examination of touch might come across as sentimental or squishy. I won't command you to breastfeed your baby or hug your coworkers. Our reticence about it is natural, and it's important to respect each other's comfort levels. Nor will I argue that we all need to abandon our smartphones and tablets and go back to meeting in person and writing letters on paper. But this book will make you appreciate what it means to inhabit a body wrapped up in a skin. You will think about your sense of touch differently, no longer passively and blindly but instead as a valuable, intuitive tool. It's time to consider the consequences of a culture that appreciates seeing over feeling and, by extension, thinking over being, and whether these are values we truly want to carry forward.

Touch gives us our grasp on the world. It's the language of our inner lives.

HOW TO FEEL

1

Dull

How Our Cultures Lost Touch

Karl Marx inhabited a unique tactile world. He was born in 1818 and grew up in a middle-class home in Trier, a town that is now part of Germany. His father was a lawyer who was active in liberal politics and the first in his family to receive a secular education after a long line of rabbis. On the side, he cultivated wine, which was considered fashionable in his community. He had built his family a comfortable, middle-class life and hoped his son would follow in his professional footsteps. But in college the younger Marx veered in another direction. He was interested in philosophy and literature, not the law, and more than that, he liked getting into trouble. He became the copresident of a drinking society and even got into a duel, and his grades deteriorated. After switching schools, he eventually managed to obtain his doctorate.

Although he hoped to enter academia afterward, he was denied a position because his ideas were so radical. So he took another turn his father disliked and became a journalist. The work suited him well enough, but it hardly paid him anything, which meant he would never again enjoy most of the luxuries he grew up with. After jumping around to projects in several cities, he and his wife, Jenny, his sweetheart since childhood, settled long term into one of the poorest neighborhoods in London and lived humbly, to say the least. They didn't have real furniture in their house. There was a chair with a leg missing, and the sofa was tattered. They lived in a state of general filth, and there were some periods when they didn't have enough money to eat.

Marx's life was a mess, as was he. His hair was unkempt. His clothes were rumpled. He often worked surrounded by chipped teacups, dirty spoons, his children's toys, and piles and piles of books. He wrote all night in puffy clouds of tobacco smoke. His surroundings reflected his internal agitation. He believed these circumstances were the price he had to pay to pursue the work he was meant to do. For relief, he nursed his vices, including cheap cigars, red wine, and decadent foods, all of which took a toll on his physical health. His senses were under constant assault.[1] This was certainly a departure from the clean, ordered expectations of his childhood. It must have been even more jarring for his wife, who grew up better off than he did.

His poor habits caused him to develop liver and gall problems, which gave him a discomfort within his own body. The most visible manifestation of his illness were the painful boils that erupted all over his skin. Some experts have retroactively diagnosed him with a condition called hidradenitis suppurativa, in which certain sweat glands, found mostly around the armpits and the groin, become blocked.[2] When these lesions, which he called carbuncles, rubbed against his pants, he was in so much pain that he could hardly take a short walk. Sometimes he couldn't even sit down. He was repulsed by himself. When he was particularly irritated, he attempted to shave the boils off one by one with a razor. He admitted that the state of his body affected the tone of his prose. "The bourgeoisie will remember my carbuncles until their dying day," Marx cheekily wrote his friend Friedrich Engels, with whom he wrote the *Communist Manifesto* in 1848.

Marx was a sensitive man. He adored his wife and children, even though he made them suffer economically and otherwise. For all of their difficulties, they were loyal to him in return.[3] This must have made their shared living condition all the more painful for him, and perhaps because of that he spared them the worst of his impulses and poured his vitriol into his writing. Some historians believe his litany of health problems, beyond the eruption on his skin, could have contributed to the biting character of his prose since problems of the liver are often associated with foul moods and rude, cruel behavior.[4] It isn't that much of a stretch to say that Marx's physical alienation from his own body is evident in his interpretations of his time period and in his work.[5]

In his "Economic and Philosophic Manuscripts of 1844," a series of notes that were published only after he died, he writes about the ways the proletariat's senses were deadened by the degrading nature of their work. He was mired in "the *sewage* of civilization," he wrote, living in cramped quarters, exposed to stench and noise and a blinding haze of dust in the streets, and cowed by the deafening noise and fear of bodily harm from machines when locked away from the elements in the factories. This constant assault and lack of exposure to finer sensory pleasures rendered him incapable of appreciating them. The debased worker "only feels himself freely active in his animal functions—eating, drinking, procreating."[6]

In contrast, members of the bourgeoisie should have had the money and freedom for more meaningful sensory pursuits, but they were trapped in a different way. The bourgeois business executives had abandoned the pleasures of sensation in favor of what always took precedence in a capitalist society: the accumulation of wealth. In fact, there was a kind of pride that came from self-denial and self-sacrifice that this endeavor required. "The less you eat, drink and read books; the less you go to the theater, the dance hall, the public-house; the less you . . . sing, paint, fence, etc., the more you save—the greater becomes your treasure which neither moths nor dust will devour—your capital."[7] If people of any class were going to come into themselves and redevelop their sensory refinement, an eye for beauty, and an ear for music, Marx wrote, they had to be freed from their numbing economic predicament.

Marx may have been using his own life to describe the feelings of the masses, but years after he died, scholars of history, sociology, and anthropology realized that he had intuited something important when he wrote "the *forming of the five senses* is a labour of the entire history of the world down to the present."[8] His was one of several texts that influenced the humanities to start noticing that our senses, including touch, are worthy material for personal and social critique. Since the 1980s, these fields have been slowly correcting for their primary focus on visual culture, such as paintings and text, which has been limiting and overlooks how the other senses play into our experience.[9] Just like Marx, we're all making important decisions about ourselves and our place in the world based on how our bodies feel, which is a product of time, place, and identity.

David Howes, an anthropologist and director of the Concordia University Centre for Sensory Studies in Montreal, shared Marx's story with me when I met with him to talk about how we experience touch day to day. The famous philosopher and political theorist explicitly stated something that we all tend to do subconsciously. We each inhabit our own personal sensescape that shapes our personalities and opinions, and yet most of us don't recognize how our sight, hearing, touch, taste, and smell experiences are loaded with baggage from the past. We move through our lives mostly unaware of how exactly we're taking in our surroundings, especially through touch, but there are techniques for stepping outside of our usual way of perceiving the world and noticing what's going on underneath.

Howes describes his approach, which he calls sensory ethnography, as grounded in "participant sensation," or "sensing and making sense together with others."[10] If we make ourselves more aware of the aesthetics of the senses and how history and culture have shaped how we use them, then we can notice how they're influencing our interactions. To initiate me in sensory ethnography, we go for a "sensewalk" in the vicinity of Concordia University's downtown campus. When we step out of his building, he tells me that usually when we're walking through an urban landscape we pay attention to what we see and hear. But today he wants me to focus on what I can sense through my skin, both what I can touch and how I feel within my body, and to consider the values I'm gleaning.

"The skin of the city is the surface of the buildings," says Howes, who honed his method over decades of research exploring the sensory qualities of different environments and cultures in places like Papua New Guinea and Northwestern Argentina. "Just as you can tell a lot about a person from their wrinkles, from all of the little marks and scars on their skin, so the city has this residue of history on its surfaces."

We start out in an area with tall modular towers from the 1980s, and the first thing Howes wants me to pay attention to is the texture I see. The buildings still in their original form have rough concrete surfaces, which Howes has me rub with my hands. There are others nearby that have since added modern facades, and their texture is entirely different—ceramic and so smooth that it's impossible to guess how old they are. Over time, architects have moved toward sleeker designs, which are visually appealing but aren't as interesting to touch. These aesthetic upgrades signal a shift in our sensory values, toward admiring our built environment from a distance rather than interacting with it directly. We approach an even newer building covered in reflective glass panels that takes this concept to the extreme. To touch it would leave a blemish on its shiny surface.

Obviously, we don't make an everyday practice of putting our hands all over buildings as we pass them. But the reason it can be useful to experience them in this up-close way is to remind ourselves how much tactile information we're taking in from them, even without touching them at all. We take a detour into an alleyway where construction equipment buzzes loudly. The walls here are made of cement, but they're pockmarked and jagged, coarser against the fingers than the university buildings. But even without physical contact, we already know the textures are distinct in places like this; that's why we describe them as "gritty." From a distance we still interpret information in tactile terms. Our senses are constantly moving outward, away from our bodies, and mingling with what's around us. They are not passive receptors, Howes insists.

We sit down on a bench in a pedestrian plaza while admiring the fall's yellowing leaves. There are piles of leaves raked up into piles on the sidewalk, and we reminisce about what it had been like to jump in them as a child. As adults, we don't interact with our surroundings in the same way. We learn that behaving with dignity requires moving through the world at a cool remove. Howes suggests we sit directly on the ground, but we realize that it would get in the way of people who are walking around. So, instead, he invites me to imagine what I might notice if I tried it out. He points out the shapes of the paving stones on the ground, which would be interesting to feel if we were to lie down on them. When we're up high, we can't fully appreciate them.

Then he gestures at the friendly looking polka-dot shape of the metal grates that surround the tree trunks. They're there to protect the trees but also to prevent the homeless from lying down in the shade, he says. Even the benches in this plaza are carefully designed with armrests that prevent anyone from using them as a bed. Most people passing by don't notice what they are for because our immediate instinct is to sit upright on top of the bench, as Howes and I do. We don't think about how someone else might want to use it to recline on. After asking me to consider how hostile this setting is to those who are most vulnerable, he has me think about how it might be used in a different culture. In some cultures people would prefer to sit cross-legged on the ground, and in others to squat on top of the bench.

"How we're sitting reflects our alienation from the earth," Howes says. "A sensory ethnography means to sense along with the people being studied so as to arrive at an empathic understanding of their world. We often don't appreciate just how unique or different our ways of sensing are from those around us."

Much of Howes's work involves putting himself in the minds of different types of people and imagining what it's like to be in their bodies. We're always habituating ourselves to our environment, so mostly we're unaware of what our senses are up to. We have to take ourselves out of this rut to gain a more complete perspective. Howes first had such an experience in the 1990s in Papua New Guinea, where he conducted his first research on the anthropology of the senses.[11] He went in with a very simplistic idea—that literate cultures are more eye-minded due to their reliance on writing or print, while oral cultures, whose dominant medium of communication is speech, are more ear-minded. But he came away with an understanding of the senses that was much more complex.

In one region called the Massim, off the southeastern tip of New Guinea, which is known for its volcanic islands and wide expanses of sea, there's an emphasis on the importance of hearing. The chief god has enlarged ears, and children are told to listen or else they'll go mad. But Howes also found that the Massim people have a different understanding of the sense of sound than we do. It's not just about passively taking in noises but also producing them. They value hearing themselves speak and being spoken about by others. Massim people believe that to have a mouth is to be civilized. It's the way we create language and appreciate food that brings community together. There's even a story about the ancestors of humanity, who lacked mouths. They only became human once they developed them.

In another region, called the Middle Sepik, named after its location on the country's largest river, vision ranks foremost in the sensory order. There a person's social status is determined not by their notoriety (or *butu*, a term that means both "noise" and fame") as in the Massim but by how far they can see. In a harvest ceremony each year, senior men congregate within the confines of a sacred house and create noise with flutes, bullroarers, and other instruments that are meant to imitate the voice of the spirits. Women and those who aren't initiated can only hear from the other side of the woven screen that encircles the men's house, so they don't know the source of the sound. The argument, then, is that men's unique view of the origins of this commotion—since they are the only ones permitted to see the instruments—makes them superior. Interestingly, there's no assumption that they have any superior ability to sense. Rather, it's through ritual that they are able to assert control.

But in their case, the dominant sense doesn't tell the entire story. Whereas men are the makers of the bark paintings of spirits that adorn the ceiling of the men's house and are recognized as the masters of all they can see, women are expert basket-makers and weavers, a sign that they have special powers involving touch. This ability is symbolic of how they knit the community together, while men are the cause of separation. Howes says that we can't just look at a single part of a culture's sensory landscape. We have to study it as a whole. It's important to see how each one defines what the senses are and how they interact with each other. There are as many psychologies of perception as there are people, and these variations change the way we think.

Not everyone in the world, for example, even agrees that there are just five senses. The Hausa of Nigeria, one of that country's largest ethnic groups, recognize only two senses: visual perception and nonvisual perception, which includes thoughts and emotions. This division speaks to their belief that, for the most part, the senses don't act independently but rather they constantly influence each other.[12] In classical Indian philosophy there are eight senses—the breathing organ, the speech organ, the tongue for tasting, the eye, the ear, the inner or thinking organ, the hands for working, and for touch. The inclusion of thought as a sense indicates a close relationship between mind and body.

Tactile preferences aren't universal either. Howes found that in the Massim region a premium is attached to smooth sensations: well-oiled skin or the highly polished seashells used in ceremonial exchanges. They sail forth to engage in ceremonial exchange on distant islands. The preferred dance style involves fluid, bobbing motions, which emulate the experience of cresting over the waves in a dug-out canoe. This is in marked contrast to the Middle Sepik, where hardness is valued. The Kwoma people of the Middle Sepik River region derive their livelihood from the swamps, they build their houses on the ridges, and they decorate their skin with ridges through a practice called cicatrization. The hard bumps are considered pleasing to the touch and are seen as a sign of fortitude. Having a hard skin is important to the Kwoma because they are always wary of being attacked by outsiders. Significantly, their dance style, which consists of stomping around in a circle, is also geared to the assertion of boundaries.

Our senses aren't just passively taking in what's around us. We hold beliefs about them that influence how we create our art and rituals. As Howes and I, looking for a relief from the cold, make our way back to his office, I notice the feelings of the city more—the brisk wind, the surfaces of the walls, the repetitive beat of my boots on the ground. I put myself in the bodies of other people around me. I think about what it would be like to walk through the streets as a man or someone from another ethnic group or with another sexual orientation. An immigrant could have a completely different sensory hierarchy than someone who grew up here. Our identity and orientation toward the senses determine what it is that we feel around us.[13]

I remember a specific moment during my childhood when I felt viscerally the uniqueness of the way my surroundings felt to me. I was returning to my home in Sacramento after spending four months in India with my family when I was in sixth grade. Exhausted from jet lag after a long flight, I kicked off my shoes and was about to run straight to bed. I had to pause, though, when I felt my feet land on the carpet. It was destabilizing. It felt like I was sinking into the ground. When I made it upstairs, my bed was so soft it was like I was swimming in tub full of lotion. I hadn't really noticed how hard and tough everything felt when I was in India, where I often slept on straw mat on the hard ground and the bedding and pillows felt like sandbags instead of puffy clouds. But coming back, the differences were striking.

I made mental notes for the next couple days. There was the gentle spray of the shower, instead of the bucket and mug I was used to, and the ever-present air conditioning. It wasn't long before these small comforts faded back into the background. But months later, I was still cataloging all the interpersonal differences. When I stayed over at my friends' places, I noticed the way they closed their rooms at night to sleep by themselves. In my own home in Sacramento, I slept in a bed sandwiched between my parents. I draped my arms and legs over their bodies at night. I started to feel self-conscious about it, even though for my cousins back in India, there was no cutoff when it became inappropriate to share a room with parents, if not the same bed. Even when they were much older, my cousins' mothers every weekend massaged their scalps with coconut oil, a ritual that I already felt too old for.

There were some ways my family members in India were more physically affectionate than others I saw in the United States. But in other ways, they held themselves back. While adults I observed in the States hugged upon seeing each other, my relatives in India just said hello or bowed their heads. Even couples didn't touch much. My own parents kept a physical distance between themselves. I've never seen them kiss or sit on the couch draped over each other or even hold hands. It has always been a mystery to me how they interact when they're alone. At friends' homes, I saw their parents kiss perfunctorily in the same room as their kids, which I thought was incredibly scandalous.

Of course, my sample size was small. One academic who has looked into how different cultures form community is Peter Andersen, a professor emeritus of communication at San Diego State University. Andersen conducted studies on intercultural nonverbal behavior in public places such as airports back in the 1980s, when loved ones could say goodbye to each other at the terminal, before terrorist threats required heightened security. He noticed marked differences in the behaviors of various ethnic groups. As he tried to determine the reasons why certain cultures seemed to have a proclivity for sensual expression, he came up with two broad factors.

One is temperature. People from Arab countries, Central and South America, the Mediterranean, and Southern and Eastern Europe use touch the most freely, while people from colder locales are more withholding. Andersen speculates that northern people may be more serious-minded because historically they had to do more planning to get themselves through the barren winter months. The weather made them by nature more proactive and productive but also slightly chilly. Because people from warmer places didn't face these pressures, they developed cultures that are more relaxed and open. It lets them wear clothing that exposes more skin, removing one barrier against touching. Sunlight may also fire up the endocrine system, which is why animals get frisky in the springtime.

The most withholding countries aren't necessarily the coldest. They're scattered across a variety of climes and include places such as Japan and South Korea and South Asian countries such as India. This is because of the second factor Andersen proposes: their level of individuality versus collectivism. "In collectivistic cultures, individual impulses to touch defer to the collective who may find touch unseemly or impolite, a departure from the shared norms that regulate individual impulses to preserve group harmony," Andersen writes in *The Handbook of Touch: Neuroscience, Behavioral, and Health Perspectives*, a collection of academic writing.[14]

The reason collectivism is common in Asia, he says, is because the continent has some of the oldest continuous cultures in the world, Andersen says, so there has been a long period for unspoken rules of decorum to take root. The way they may have maintained peace for so long is in part because of their habits around touch. Avoiding touch in public is a way to emphasize the entire group's comfort above the closeness of any individual members. Religious tenets about touch are another reason certain cultures might avoid it. Meanwhile, cultures that emphasize the individual, which is often the case for younger countries, may reward confidence and expressiveness in personal relationships.

In my own family, I saw this theory in action. When the family is a unit that consists of great-aunts and third cousins, the individual household is subsumed within a larger unit. The allegiances are held together through custom and ritual, not physical closeness between individuals. Many marriages, for example, are arranged by elders and emphasize joining families rather than couples. A wife is traditionally expected to go live with her new husband's family. Given these dynamics, it only makes sense that the couple's romance wouldn't be outwardly expressed. Instead, the entire community's affection gets poured onto the children they have together.

But even as I make these observations, it's also important to acknowledge that no culture is uniform. There are plenty of variations even within families, and people and cultures are capable of changing over time. My own aunt, when she learned I was writing a book about touch, emailed me unprompted about the lack of physical affection between members of our family. It's something she probably wouldn't have noticed had we not been somewhat unusual even amid our larger cultural context. She had an arranged marriage to an Indian man, and she envied the casual loving behavior she witnessed in his family, particularly between the women.

"I have always felt we have not much used the sense of touch as a form of communication within the family to convey our mood," she says. "Our parents have not hugged us even once either to congratulate us on our achievements or console us when in sorrow. I am not able to say it is right or wrong. That is the way we have been brought up. I have noticed my mother-in-law and sister-in-law used to hold one's hand and talk. I feel in that way they drive their point more powerfully."

I'm not sure how to consider our lack of touch either. Part of me thinks it would have been strange for us to touch more. None of us has an especially outgoing personality, so it wouldn't come naturally for us to be more affectionate. The way we touch each other is often an expression of our inherent introversion or extroversion. But it's interesting to know that my aunt has the desire—or at least an appreciation—for another way of being. Maybe we all experience a push and a pull with touch. Our culture instructs us not to do the things our skin and our bodies crave the most.

Where does the United States stand in all of this? Well, there are mixed opinions. Many people label the United States as a low-touch culture, possibly because of the influence of nonexpressive British immigrants who founded the country. But Andersen isn't so sure. For him, because we have brought in tactile traditions from all over the world since the country's early years, there's a heterogeneity to how Americans touch. It's neither a high- nor low-touch culture. Everyone is different, which has made us all have to be more careful about navigating everyone's various nonverbal cues and surely gives us more awkward encounters to laugh about.

Culture isn't the only lens through which to examine how we use the senses. We can also see how our understanding of the senses has shifted over time. Just think about how the past must have felt in the West. In the Middle Ages, life was rougher, literally. A laborer would have gotten up early to the sound of a rooster crowing, pulled on a set of scratchy woolens, and headed out for a day in the unforgiving elements. Embedded in these feelings were critical messages that affected what they valued, which then helped them shape the tactile world they proceeded to create for themselves. In her book *The Deepest Sense,* cultural historian Constance Classen, a collaborator of Howes at Concordia University, dives into the sensory turn that took place during the transition from the Middles Ages to the Enlightenment to the Industrial Revolution. She writes,

> The very use of the term "Dark Ages" to refer both to the centuries immediately following the fall of Rome and to the entire medieval period conveyed the notion of an age when people groped about blindly, feeling their way through life. Indeed, according to this sensory classification of historical periods, it was only in the Enlightenment, the eighteenth century age of reason, that the light of learning finally dispelled the shadows of past ignorance and enabled people to think clearly about the world.[15]

In the Middle Ages, people's value came from their mastery of their bodies and braving harsh conditions. A typical laborer spent long days planting, harvesting, and slaughtering, developing their skill through daylong repetition. They cultivated empathy for the land and animals under their care. This allowed them to read the signs from nature about how to use their time and resources to reap the most rewards. Moving up through the rungs of society were blacksmiths, cooks, and servants who all developed similar physical skills. Even doctors had to brave the weather to make house calls and diagnose patients using manual abilities, such as palpating and pulse reading. At the highest rungs of society were knights, who had some of the most strenuous jobs of all.

Women of a certain class were able to avoid much of this daily hardship. In exchange, they were expected to provide comfort to their families. A fifteenth-century housekeeping manual called *The Goodman of Paris* described some of the pleasures a wife might provide her husband after a long tiresome day at work: white sheets, furs, and other joys. Women weren't typically the recipients of this kind of treatment, but the arrangement worked out for them in some respects because they could avoid some of the tougher work. However, they did become symbolically linked with the lowest sense, which was used as an argument to justify their subordinate position. The softness and sensitivity expected of them was not prized as much as the harsh labor of their spouses.

The physicality of work was mirrored in people's social lives. They ate with their hands out of shared bowls, slept in familial beds, bathed together in large tubs, and spent their evenings huddled around a toasty hearth. The warmth of being around friends and family wasn't just metaphorical. It was devised intentionally by ritual. Each body belonged to the group, and its purpose was to support its collective health and cohesion; there wasn't room for people to have boundaries. Embracing and kissing wasn't limited to intimate relationships; people kissed others of the same gender platonically. Touching took on a symbolic role in business dealings. The handshake, a greeting that dates back to at least the fifth century BC, represented a binding contractual obligation.

Even spirituality was more sensual. Saints were thought to have a magic touch, which is why churches practiced the custom of the laying on of hands during baptisms, confirmations, healing services, and blessings. Artifacts of deceased saints, such as old bones, traveled from village to village, and people derived satisfaction from being in contact with them, sometimes by placing them directly in their mouths. But the signs of a pivot to a new set of sensory values came from the church as well. Many leaders considered the soul to be under constant attack by the vices that entered the body through the senses. All worldly indulgences, especially sexual ones, were considered sinful. The message was that the body's desires, including for touch, needed to be kept in check and that a superior form of spirituality could be achieved by encouraging people to reflect on God purely with their minds.

The eye had long been associated with the intellect, so it became a dominant sense as the literacy rate rose at the dawn of the Enlightenment. Seeing overshadowed doing as the way people proclaimed their expertise. They no longer had to spend years apprenticing under the supervision of an expert, carefully training the intuition of their bodies. Instead, they read and took tests and received diplomas from prestigious houses of learning. Medicine was one field where this trend was especially observable. Folk healers and midwives, who had developed over the course of their lives an intimate bedside manner, were phased out in favor of professionals with more formal knowledge.

As newly educated people moved to cities where their social ties became weaker, their interpersonal etiquette emphasized maintaining a healthy personal space. The new rules required imagining each person's body was wrapped in an invisible bubble to avoid any unnecessary discomfort. When people greeted each other, they did so from a distance, a subtle show of respect for another person's autonomy. When dining, they sat in their own chairs with separate place settings and had to transfer food from communal bowls onto their own plates before eating it. In an environment filled with strangers, people could no longer trust a handshake deal, so they started drawing up enforceable contracts, a practice that grew steadily from the 1600s on.

City dwellers were the first to adopt these manners, which created an obvious culture gap between them and those from rural settings. They became a signifier of social class. Among the urban elite, touch even became uncommon among family members. In France an entire literary genre centered around men writing letters to their sons to express all the heartfelt emotions that they repressed in their daily lives. It was as if, without practice, they forgot how to express affection for even those closest to them. As social touch of all kinds became less frequent, every little brush against someone else's skin took outsized significance. In literature, aristocratic protagonists were immensely sensitive. Small gestures and caresses played on their nerves like a bow on violin strings.

Religious services became even more hands-off. Instead of bowing or clawing at relics, churchgoers meditated privately on images of God. Even the notion of hell was shifting. Previously, hell had always been described as a painful place of fire and pitchforks. But it was transforming instead into a site of mental anguish where sinners would spend eternity ruminating on their failures in life. Museums began to impose rules against touching artwork, in part for the sake of preservation but also to establish themselves as spaces to be revered.[16] Politeness meant looking, not touching. This directive slipped into a more serious belief that admiring from a distance was a superior way of appreciating beauty. Painting came to dominate the high arts. Pottery and weaving, work that was better suited for exploration with the hands, were demoted to mere craft.

The concept of a sensory hierarchy even made its way into how people from other races were perceived, which was really just a way to justify the existing racist social order. Lorenz Oken, a natural historian, drew up a sensory scale with the European eye man, whose predominant mode for learning about the world was seeing, at the top. By suppressing his baser senses, the European man established an air of respectability and refinement. Asians were thought to be ear men. Because the ear is also a distance sense, this meant Asians had the second-highest rank. Then there was the Native American nose man and the Australian tongue man. Cultures whose prime mode of sensation was touch, represented by the African skin man, were thought to be the least developed.

In the Industrial Age, machines such as the power loom and the steam engine could produce goods much faster and with more precision than people could with their own hands. This meant that physical labor was demoted yet again. People who had once prided themselves on their manual competence were consigned to an assembly line where they became anonymous cogs. They were expected to operate as machines themselves. Their bodies remained in constant repetitive motion. Work by the upper classes became even more mental. The automobile allowed people to see so many sights so much faster than ever before, and that montage became an apt symbol for the pace of modern life.

"His brain is a racetrack around which jumbled thoughts and sensations roar past at 60 miles an hour, always at full throttle. Speed governs his life: he drives like the wind, thinks like the wind, makes love like the wind, lives a whirlwind existence. Life comes hurtling at him and buffeting him from every direction, as in a mad cavalry charge, only to melt flickeringly away like a film," writes Octave Mirbeau in his 1907 novel *La 628-E 8*, titled for the license plate of a car used on a road trip that creates a rapid succession of scenes meant to symbolize the human psyche.[17]

As average living standards improved, people of all classes were better able to afford the kinds of comforts that were once only available to the elite. They bought glass windows to keep out the cold and the bugs. They filled their homes with smooth textiles, soft beds, and plush couches. Suffering was no longer seen as a sign of strength, and the pursuit of sensory pleasure was no longer considered sinful or weak. As all of life became more comfortable and uniform, people started to rethink beliefs in areas as diverse as child-rearing and corporal punishment. Spanking, which was once the norm for parenting, began to be considered harmful. So was too much affection. In prisons, order was maintained not through punishment but through the use of surveillance cameras and physical isolation. These changes were believed to be more humane, but they were punishing in new ways.

Our sensory experiences have always shaped the times we live in. Working with a hammer or a wooden hand tool versus heavy machinery, wearing a billowy gown with a waist-cinching corset instead of a sheath dress, or greeting someone with a curtsy rather than a hug can impart subliminal values. These feelings contribute to our beliefs about the nature of work, our relationship with our bodies, the ways we form connections within our social groups, and the differences between the classes. Sometimes these implicit messages about the senses tell us how we how interpret touch, as knowledgeable or overly sentimental, as friendly or uncouth. And the fact that today, surrounded by indoor comforts, we hardly notice our sense of touch at all continues to mold our concept of our senses.

Our sensory evolution didn't stop after the Industrial Age. The digital era has brought new developments that have upended the way we use our senses yet again. Much of our face-to-face interaction has disappeared. We spend less time talking and more time typing. We're surrounded by screens and speakers, and any sense that isn't vision or hearing has taken a back seat. Even the way we experience the so-called lower senses is now determined by how they look to our eyes. We choose where to eat based on how photogenic the food is. A get-together with a friend doesn't carry much consequence unless it's visible on social media. Technology has had a homogenizing effect on sensation everywhere.

These same changes are happening around the world, in places with vastly different sensory hierarchies, and it's even evident in my family members in India. Office jobs in the tech sector have led to less walking and movement. New appliances make cleaning and cooking less difficult. With more wealth, my family has built private bedrooms and bought soft beds and couches and air conditioners. A younger generation is choosing to marry for love, which has led to entirely new ways of relating within a larger family context. With those closest to them, they've become slightly more demonstrative, but, as is the case here, the majority of communication happens at a distance, through text messaging on WhatsApp.

We've all, everywhere, had mixed feelings about a technological world that is lacking in physicality. While we enjoy the convenience of our devices and for the most part admire their sleek appearance, we also wonder if there's something missing when we can't feel ourselves interacting with our technology or with each other. These worries will either get buried as we continue down the path we're on and adapt better to new, visually based forms of interaction. Or we could decide that technology is not serving us and either revert back to older, simpler ways of connecting or find ways to incorporate more of our senses into our technology to re-create the feelings of real life.

In the evening, after my day with Howes, I test out my new analytical powers with his Ph.D. student Erin Lynch, who is doing some research on the sensory aspects of the Casino de Montreal, a short cab ride away on Ile Notre-Dame. It's a place that brings to the fore how technology is changing the life of our senses. When we arrive, the jagged exterior is lit up, like a paler version of Emerald City from *The Wizard of Oz*. It's a special Halloween-themed night. Actors dressed as werewolves from the "Thriller" video meet us at the front doors, and there are monsters parading through the halls. But apart from this spectacle, it's about what might be expected at a state-run gambling locale. People are set up at slot machines, which are all lit up, pushing buttons with a glazed look in their eyes.

Lynch has me analyze what I'm taking in. I tell her that I noticed some of the older machines have mechanical wheels and hand cranks. They feel a lot more fun to play, and something about the tangibility of the moving tiles makes it seem harder for the house to cheat than a digital display, which I know isn't true at all. The newer machines are all about giant screens that are tall and curve over the user so the user is fully enveloped. The machines have as few moving parts as possible, with extremely smooth buttons that take very little movement to push, which to me makes them less friendly and engaging. But—just like the buildings from earlier—they're indicative of society's move toward the visual.

Lynch says she suspects the design is actually intentional. The idea is to provide as little resistance, physical and emotional, to the user as possible. It's about getting people to lose track of their bodies and find themselves lost within the game. If there were moving parts, they would feel the fatigue over time, a sign for them to stop. The lack of touch allows the process of gambling to become dematerialized, along with the concept of money. That's why these machines also don't spit out real tokens anymore either. They just display a number that has to be printed out and redeemed for actual cash. Any tactile effects that do exist, such as ergonomic chairs that rumble in excitement after a winning spin, are made to keep them sitting there longer.[18]

Because of the Halloween party going on, there's a loud band playing "Celebration." But if it were quieter, she says, we could hear the machines pinging after someone's big win, which would add to the communal and competitive aspect to what's usually a solo activity. Over at the table games, there is a totally different environment. Everyone is hushed. The dealers are wearing all black, and the lighting is low. Even though we're standing at the roulette table, which is a game of luck, people's subtle expressions indicate they're playing it as a game of skill. The way they hand out cards and collect the chips is elegant and practiced. Here, it's the subtleties that are important, and touch is highlighted.

This scene goes to show that while technology is evolving and incorporating even less touch, we still prefer it in some areas. The casino experience is symbolic of how, just as in the past, we continue to simultaneously marginalize touch and celebrate it. My sensory ethnography has let me see that my surroundings are constructed based on cultural understandings of the senses, and they tell me wordlessly how I should be interacting with them; depending on who I am, they're either welcoming or forbidding. I'm not even aware when they're guiding me to have certain beliefs about my place in the world, about what's polite and what's modern.

The story of touch, and where it could go in the future, is likely to have more twists and turns than we currently understand. But before looking forward to what we can expect, it's important to consider what touch is to start with. Touch is difficult to understand because it's unlike any of the other senses. There's no single locus of sensation, like the eyes or the ears. Instead, it's spread all over the surface of our bodies, and even inside it. And there's no obvious way to define its unique purpose—what it tells us that the rest of the senses aren't able to. As we'll explore next, it's an incredibly complex sense that uses multiple abilities of our bodies working together.

2

Numb

Life Without Touch

At nineteen years old, Ian Waterman moved from his family's home in southern England to Jersey, an island in the English Channel, to take a job at an artisan butcher shop. After quitting high school and bobbing around for a few aimless months, he was hopeful this opportunity would lead to a new, long-term career. He paid careful attention as his boss, who prided himself of his French training, taught him the art of boning short ribs, butterflying legs of lamb, and making crown roasts of pork. Waterman practiced diligently and worked extra hours to fulfill the endless orders. He regularly won friendly competitions with his coworkers over who could prepare their cuts the fastest. He impressed everyone he worked with, and soon he was offered a role as manager.

But before the contract was ever signed, Waterman woke up with what seemed to be the flu. He had an upset stomach and fever, and his body ached. He tried to go in and fulfill his duties as usual, but it was impossible for him to keep up. A few hours in, he was so exhausted and weak he couldn't even hold a cup of tea. Still he continued to press on. That was, until a coworker finally suggested that he go home. The workaholic in Waterman wanted to protest, but he also recognized how awful he felt, so he quietly complied. After a few days in bed trying to fight off his illness, things got much, much worse.

The proprietor of the hotel where he was staying took pity on Waterman and offered to clean his room if he would get up. She knew he was young and living by himself for the first time. But as he tried to get out of bed for her, he collapsed like a wet noodle. She asked him if he was drunk. "No, you've seen me drunk," he said. She had; on nights off, Waterman often enjoyed long dinners over many glasses of wine with his new foodie friends. After a lonely childhood, he was delighted to have built a strong community around him. "I know what drunk is. This isn't it." She suspected there was something seriously wrong and called for an ambulance.

After his first night at the hospital, Waterman woke up with a hand around his neck. He thought he was being strangled and started hyperventilating. But as he raised his head to check who the perpetrator was, he saw that it was himself. It was his hand. He just couldn't feel it. Once over that shock, he took stock and realized that it was like his entire body from his neck down had gone missing. He couldn't sense his back pushed up against the hospital mattress, which gave him the impression that he was floating in air. The only way he could tell he existed was by looking down at his body. When he told doctors about the problem, they were perplexed. They pricked him with pins and tapped him with patella hammers, neither of which he could detect, and they could come up with no answers. They decided to wait and see if things would improve on their own.

Several weeks later, after many more tests, nothing had changed. However, the doctors did develop a theory for what had happened: when Waterman came down with his illness, his immune system overreacted and killed not just the virus but also most of his somatosensory nerves, a complex system that responds to a whole host of changes to the surface and inside of the body. The specific nerves that were affected included those that transmit information about feelings such as pressure, stretch, and vibration on the surface of his skin as well as those responsible for registering body position, the so-called sixth sense of proprioception. Only his pain and temperature sensors continued to function normally. He is one of only ten known cases of this nameless disorder.

When we talk about touch, we often think about just the information we get through the body's surface. But touch actually involves many parts of our anatomy working together. To experience feeling through our bodies, the skin has to be moving across a surface, and there needs to be some inherent knowledge about that motion. In other words, we need three things: the sensation emanating from the skin that we colloquially call touch, the capacity for movement, and proprioception or our knowledge of where our bodies stand in space. This combined ability is referred to as either active touch or, more often, haptic perception. Waterman was missing two of its three components, which really meant none of them were fully functional. Although he was still capable of movement, he had no awareness of it. Waterman first fell sick in 1972. Today he's in his sixties, and he's never recovered.

Ask people which sense they would least like to lose and most will say vision. But I suspect that's because we don't think of touch as something that can go away. Closing our eyes gives us some idea of what blindness is, colds regularly down our sense of smell, and noise-canceling headphones give us something close to hearing loss. The closest thing people can come up with to losing touch is paralysis. But that also involves a loss of movement. It's hard to relate to what missing our ability to touch entails. That's what makes Waterman's case so fascinating. By living without the kind of bodily knowledge that most of us take for granted, he is able to show us what the tactile experience even is.

As my trip to England to meet him nears, there are dozens of questions circling in my mind. When his eyes are closed, how can he know if he's sitting down or standing up? When out for a walk, would he even be able tell if he was bitten by something? When he's at the grocery store, how does he judge whether the lemons are ripe if he can't feel them squishing under his fingers? When my foot goes numb, am I getting the localized experience of what he feels all over? Or is it more like trying to operate something while wearing a thick pair of gloves? What do hugs mean to him if he can't feel them? And, well, what does sex feel like?

Waterman is a tall man with a sloping forehead and light bulb–shaped nose, out of which sprouts a thick brush of mustache that nearly covers his mouth. He's gregarious and funny and seems like an extrovert, but he tells me he prefers to spend most of his time in the privacy of his secluded bungalow. Waterman lives just off England's Jurassic Coast, named for the remnants of the ichthyosaur, a swimming dinosaur, once excavated in the area's bleached cliffs. It's a place where local families go on summer holiday, where grandmothers in baggy dresses chase around children looking for ammonites. His neighborhood is just inland but much quieter, with roads no bigger than walking paths that are lined by tall, forbidding hedges that act as dividers between farmland that had been sliced up over the years.

Settled into his cozy living room, he walks me through how he relearned to function again without bodily feedback—a feat that should have been impossible—by using the same dedication and obsessiveness that he had once directed toward butchery. He started to teach himself even while still at the hospital. At first he attempted eating on his own. But when he lifted a spoon to his mouth, his arm flung out of control and wrapped around his head. So he slowed down, carefully concentrating on each step. First, he clutched the spoon tightly. Once he had that mastered, he bent at the elbow. Next, he moved his hand toward his face. Finally, he lifted his wrist to reach his mouth. But just as the food was making its way to its destination, his other arm rose up on its own. He sat on it to prevent it from floating away.

His next goal was to sit up. Without sensation, he had to think about each body part he was using, one at a time. He tightened his stomach muscles, but that didn't do anything. So he tried to lead with his head, and that worked better. But that only got his shoulders off the bed while the rest of him remained supine. To go the rest of the way, he had to hoist himself up by swinging his arms. It took a few tries before he could make it all the way to sitting. But even then he got so excited that he forgot to keep his abdominal muscles engaged, and he plopped right back down. After taking a short rest, he started up again.

Even as doctors saw signs of his progress, they knew they weren't going to be able to cure him. They allowed him to go home and offered him a reference to a physical therapist. But when he moved back to his childhood home, in Southampton, he became incredibly depressed. He refused to see anyone, even his brothers, bitter that he had been sidelined just as he was building a life for himself. He kept himself busy by doing puzzles alone in his room, licking his fingers to pick up the pieces. He tried to eat on his own even if that meant a meal took him hours to finish or he dribbled tea down the front of his shirt. After spending several months in this dark place, he and his mother agreed it would be best for him to stay full time at a nearby rehabilitation center called Odstock Hospital.

The specialists there took some time trying to understand his unusual condition, which they'd never confronted before. They came up with a set of repetitive exercises that involved having him make small movements with his arms and legs, with guidance from his eyes. He practiced tensing and releasing his muscles while watching how they responded. It was like learning to be a machine operator, but with his own body. He started to better understand the physics of the body and develop confidence that if he could get the mechanics right to perform a task once, he would be able to do it again. But every attempt took immense energy and focus to manipulate each individual body part.

This was nothing like the intuitive understanding the rest of us have about our position in space, but it did get him from point A to point B, which was all that mattered. Waterman had to overcome his shame that his movements would never be completely fluid and that people could see he was disabled. Through physical and psychological therapy, his mindset and coordination improved remarkably. Eventually the doctors progressed to a major step, which was having him walk.

Waterman leaned forward so he could keep all of his body parts in constant view since he needed vision to compensate for his lack of touch. To balance his weight so he wouldn't tip over, he held his arms back and tensed his back and legs. With everything in place, he attempted to move his feet. It didn't appear graceful, but with constant practice he gained confidence that if he could do it once, he could do it again. Within a year at the center, he could walk while holding onto a support bar, stand on his knees and catch balls, and write legibly. By two months after that, he took his first walk alone in his bedroom during a visit home.

"How'd you do it?" his mother shouted when she found him standing in the middle of his room, looking panicked because he didn't know how to turn around and go back to his bed. His home didn't have safety bars circling the walls, like the hospital did, and he was afraid of falling. He told her he'd walked, which he thought should have been obvious to her.

"Well that's just stupid," she said. "You're going to fall over."

She brought his wheelchair, and once he was safely seated, she dressed him down for taking such a risk, but Waterman could also tell how elated she was that he could walk again. Waterman stayed at the hospital for seventeen months in all. By the time he left, he could go out unaided for stretches. He learned to drive a car outfitted with hand controls. His handwriting was better, and he had even begun to carve out designs in small tiles of wood, a craft called marquetry. His doctors believed he had the skills to go back to living on his own again. He never regained the manual dexterity needed to cut meat. But he was able to hold a government job and live an independent life. No other known person with his condition has made that kind of recovery. But each day remains remarkably difficult for him.

It's hard to grasp what Waterman means when he talks of his struggles. The reason is that he makes it seems so easy. His gestures are only slightly stilted. If I didn't know, I would probably pass them off as a minor health problem due to aging. This has always been a circular problem for him—having a debilitating condition that makes him feel alone, and then getting so good at covering it up that people can't identify with him. Only the few others who have his condition and rare specialists, such as his neurologist, Jonathan Cole, can truly understand how hard he's working all the time. Cole has described Waterman's ability to move through the world as "a daily marathon . . . one performed on a high wire, where a fall is always possible."[1]

To try to get me to understand, Waterman asks me if I play any sports.

"I run a few times a week," I say. "I've recently taken up rock climbing."

"Brilliant. It's just like climbing. You know when you're at the bottom, and you have to chart your course to get up the wall? That's the same way it is with me. Every move I make has to be planned. It's an exercise in concentration every time, which I know sounds bizarre."

That I can picture. I've been on bouldering walls trying to figure out how to turn in just the right way to make it up to the next hold. It involves a degree of strategy that I don't need when traversing across flat ground. Waterman says he is always mentally planning out his next action in that same way, whether it is taking money out of his wallet or eating lunch or typing on his iPad. Waterman has me imagine that my limbs are anesthetized. In order to move them, I'd have to activate each one deliberately, as if with a remote control. It takes constant concentration. He mimes the way he has learned to perform fine motor tasks, such as threading together paperclips. He moves away all but the most essential fingers and rests his wrists on a table so they can't float off. Then he has to will his tiny muscles into motion.

"The other sports analogy you often hear is riding a bike. When you're learning that as a child, there's a lot of concentration, and then eventually it becomes automatic, even if you don't do it very often. But for me, it never becomes like that. I need to stay very conscious."

His wife, Brenda, a friendly, neatly coiffed woman with elegant streaks of gray, brings us cups of tea. She hands me mine. Waterman motions for her to put his down on a side table, which

I assume is because he doesn't want it right away. But then he picks it right up. He tells me that taking it from the table is easier because there are fewer moving parts. He notices me sitting perched on the edge of the sofa and observes that he could never do that. Instead, he sits firmly pressed back into his seat so he doesn't have to think about staying balanced. The less he has to focus on, the better. He is missing the ease of movement the rest of us have and is always operating at 100 percent.

His example makes me notice for the first time all the sensations that are giving me my understanding of where I stand within my surroundings. I notice the skin-level feelings, the chair pressed against my backside and my elbow slipping a little across my knee. If I pay close attention, there is also the stretch of the skin of my lower back and kneecaps. Then there's the barely there strain in my muscles that tells me I'm holding up my torso. When I get up to stand, I can feel the lengthening and contracting of my muscles and the air making contact with parts of my leg that has just become exposed. I consider how much harder I'd have to work to control my movements without these feelings.

While Waterman's disability has caused many large sweeping changes to his life, they all stem from damage to some very small nerves responsible for a few aspects of touch. To understand him—and ourselves—we have to turn to these microscopic parts. Touch isn't all about our skin, but it's still a good place to start. At an average size of about twenty square feet, our skin is often said to be the largest organ. It's not, actually. The alveoli of our lungs, which move oxygen into the blood and carbon dioxide out of it, have a far greater surface area. At almost nine pounds, the skin is not the heaviest either. Our bony skeletons, sometimes considered endocrine organs, are more than twice its weight. And that doesn't count the muscles attached to it.[2]

But the skin does act as one large, highly perceptive shell. It is covered in mechanoreceptors, sensors that detect different kinds of deformation caused by the body coming up against its environment. They act similarly to the eyes' rods and cones, which are receptors that recognize visual qualities such as brightness and color. In the same way, the mechanoreceptors are specialized for particular tactile inputs.

A few parts of the body—for instance, our hands, the soles of our feet, and our lips—are covered in what's called glabrous skin, meaning non-hairy, and it contains a characteristic set of mechanoreceptors. Close to the surface of the epidermis, the outer layer of the skin, are the bulbous, radish-like Meissner's corpuscles, which are adept at sensing light touch and low-frequency vibration, like that from a fallen hair strand or a crawling ant. Immediately beneath them, still in the epidermis, are clustered disk-shaped Merkel cells, which help to spot tiny indentations in the skin, such as from Braille letters. In illustrations, complexes of Merkel cells look something like cannellini beans strung end to end. Further into the flesh, in the dermis, are branch-like Ruffini endings in the shape of snap peas, which are talented at detecting skin stretch. Down in the subcutis, the deepest layer of the skin, are large onion-shaped Pacinian corpuscles, which act as antennas for high-frequency vibration or flutter.

In our non-glabrous or hairy skin we have even more types of mechanoreceptors called lanceolate endings, which encircle body hair, everything from the hardly visible fuzz of our faces and necks to the longer strands of our arms. Their appearance depends on what kind of follicle they surround. They recognize the flutter of each strand, telling us about the movement of air, whether it's from a gust of wind from a passing train or the speed with which we're riding a bike. So far, scientists have identified about thirty types and subtypes of mechanoreceptors throughout the skin.

In addition to these receptors for mechanical changes on the body, there are also sensors that detect other types of stimuli. For example, there are nociceptors for pain, thermoreceptors for heat and cold, and histamine receptors for itch. These senses themselves can blur between one another. For example, thermal stimulation from warmth can gradually shift to a painful nociceptive experience as the thermal stimulation reaches tissue-damaging temperatures. The density and type of all of these sensors depends on where they are located on the body. Each is highly specialized, like an individual musician in an orchestra, which means we can only recognize the piece when they're all playing in unison. It's the entire pattern of activation—the melody, the rhythm, and the harmony—that tells us what we're feeling.[3]

When we're sensing air from a fan, what our skin is actually registering is the concurrent activation of receptors for temperature, vibration, and hair movement as well as the stillness of the others. When we feel our hands resting on a hard surface, our skin is reacting to mostly skin stretch and some heat. Wetness feels so distinct that we might guess it has some kind of specialized receptor in the skin. But it isn't. This impression is created by the co-activation of light touch and temperature. The water molecules move lightly at the same time heat is being stripped away from our bodies, and that's what we associate with the movement of liquid droplets.

Once any of the touch receptors is triggered, it needs to generate an electrical signal in order to alert the brain. How this works is still a mystery for many types of mechanoreceptors, which is indicative of how touch research lags behind that for the other senses. Their receptors have, for the most part, been understood for years. But for at least some mechanoreceptors behind touch, it happens when a bend in the skin causes a pore in the cell's membranes to open, allowing ions to rush in and out, changing the electrical charge of the cell. When the electrical charge of the cell is sufficiently changed from its resting voltage by the activity of many ion channels opening at the same time, the cell produces an action potential—a rapid, all-or-nothing voltage switch. The collective action potentials of groups of cells firing together in time and space form the electrical language of the brain.

The electricity buzzes up to the brain through nerve fibers, some fast and some slow. So-called rapidly adapting fibers transport the sensations of light touch and vibration at a clip of about 250 miles per hour. The secret to their speed is a thick coating of myelin, a fatty insulating sheath. But after the initial burst, they abruptly shut off. This characteristic accounts for the sensation we feel when draping ourselves in a terry cloth bathrobe. It's cozy when it first hits our skin, but after a few seconds it's hardly noticeable. We have only a faint awareness that we're walking around clothed, which is a useful adaptation that allows us to direct our attention toward what's new.

When we're feeling pain, say from closing a drawer on a finger, it's the slightly slower A delta fibers, which move information at about forty miles per hour, that tell us we're hurt and to whip our finger out of harm's way. This is followed up by another signal from the so-called C fibers, which are some of the most leisurely, as snail-paced as two miles per hour, because they're coated with a lot less myelin. We would have to try to amble along that slowly. It's possible to count to three before we feel the second wave of pain they bring on, which is more disperse and pulsing and gets us to keep our hand protected so it can heal.

The ends of all these fibers, fast and slow, cluster together as they move up the spinal cord. The signals remain segregated in separate layers: pain and temperature in one and light touch in another. Well, they mostly stay in their own slots. There are some opportunities throughout the journey for them to swap with each other. For example, there is what's called a wide dynamic range neuron in the spinal cord that helps to integrate pain and fine-touch information, which is why our first instinct when we hurt ourselves is to press down on the injury site. The light touch cues hold off some of the pain signals on their way to being processed.

Ultimately, like all the senses, touch exists in the brain, which is where the electrical signals end up. They arrive at the thalamus, which acts as a switchboard operator. Then they are sent to the somatosensory cortex, the main touch-processing area. Located there is a map of the entire body called a homunculus, which means "little man" in Latin. There's a part of the homunculus representing our right pinky toe and our left cheek and so on. But they're not the same shape or proportion as our actual body parts. The lips, feet, and genitals are magnified, signifying those areas where we're most sensitive. Once the signal is routed to the part of the homunculus that matches where we were initially touched, we finally feel something.

At this point there is a collection of tactile inputs coming in from distinct points on the body, but they don't yet have meaning collectively. That happens in the insula, a thumb-sized part of the brain that attaches emotions to our physical experiences, including touch. It compares the tactile experience we're having to the others we've faced before, so we know how to respond. It informs us that the discomfort that we're feeling from our jeans cinching our waist is no big deal or that that a hand lotion's residue is too slippery and needs to be wiped off. The insula is the basis for our awareness of the present, and it helps drive our decisions.

But not all touch processing takes place in the insula. Another stream of information skips this step and is routed instead to a region called the posterior parietal cortex, which integrates touch with movement and proprioception. Proprioception comes through additional sensors in the tendons, muscles, and joints that are analogous to those in the skin. This part of the brain is responsible for the way we prepare to respond even before we recognize what we're touching, acting based on motor memory. It's because of the posterior parietal cortex that we might have the instinct that we're about to fall and adjust our gait or that we're about to drop our pen and hold it tighter. This is where our awareness of touch gets foggy since what happens here isn't part of our conscious thought.

Suppose we're playing basketball with a group of friends on a June afternoon. Our warmth-sensing thermoreceptors are ignited each time we step into a ray of sunshine. A friend passes us the ball. We can feel its pebbly surface via the Merkel cells on our fingertips. As we wrap our fingers around it, the webbing between them stretches, which we can tell because of our Ruffini endings. Each time we bounce the ball and it returns to our palm, we feel its vibrating surface because of the activation of our Pacinian corpuscles. All these sensors are helping us gauge the accuracy of our movements as we make our way toward the basket. As we're noticing our present action, we're also predicting what our bodies will run up against in the future. So much is happening at once, and our brains do their best to make sense of it all.

We're not born knowing how to interpret all these cues.[4] It's something we have to learn as babies, starting from our earliest experiences. Touch is the most fully developed of the senses at birth, as there's little opportunity to use senses such as hearing or vision while still in the womb. This makes it the primary way babies explore their environment. They immediately begin exploring how to read the multiple signals coming at them from their sensory neurons. Within a few weeks, most can hold their arms out toward objects in front of them, like a blanket or a toy, and put them in their mouths, where they have acute sensitivity. But newborn babies' lack of mobility still keeps their use of touch limited.

Things improve slightly when most babies are about five months old and their muscles are strong enough to let them sit up and hold their heads upright. They can use their eyes, which are now in a stable place, to guide more accurate movement. The visual and tactile systems calibrate themselves so that babies understand what's happening even when their limbs are out of sight. In this way they develop a coordinate system to interpret how they're behaving within a three-dimensional space. Eventually, they can use this system to pick up small items of different shapes and sizes. This crucial step involves many complex processes working together, including coordination between the senses and timing.

This gives them the basic visual-spatial knowledge that they can make use of when they finally stand up, which introduces new hurdles, such as monitoring their wobbly bodies and dealing with surroundings that behave unpredictably. By one year most babies have gained enough control over these moving parts that they can toddle around. This is a huge breakthrough in development. It is evidence that they can handle multiple functions at once and even predict future movement. It takes longer for humans to accomplish even more difficult tasks, such as catching a ball, coloring, dancing, playing an instrument, and driving a car. But, once learned, these skills can be commanded spontaneously because of that early period spent mastering basic bodily cues.

Throughout our lives, touch, proprioception, and movement work in a constant, automatic cycle to create efficient and controlled action and keep us physically safe. Movement leads to feelings both inside and on the exterior of the body, which then guides future movement. This haptic sense is challenging to understand because it's made up of a network of abilities. We might have separate words for the individual parts, but they come together as a single unit, which makes it impossible to understand Waterman's sensory losses on their own. When any part ceases to function, the entire system breaks down.

Waterman's condition isn't the only one that makes patients lose haptic ability. Loss of skin sensation alone is slightly more common. But because all of the parts of our haptic system flow into each other, these patients experience a similar effect. A Pennsylvania woman named Julie Malloy, who retains proprioception but has no sense of touch in her skin due to an extremely rare genetic disorder called hereditary sensory neuropathy type 2 caused by a gene mutation, told me she has many of the same problems that Waterman does because what happens on the surface of the body is so important to an understanding of body position. She has to compensate for her loss with an extra layer of effort that the rest of us don't have when she navigates through her basic routine, her work, and her most meaningful relationships.

When she is chopping vegetables, she has to think carefully about holding her knife at the proper angle and how hard to press down. She does this mostly with vision, as Waterman does. Vision tells her whether she's sitting or standing and where her legs and arms are. Keeping one eye open is what keeps her from slipping and falling in the shower. She controls her car by instinctively knowing how hard to push on the pedals and looking down at her feet when necessary. Because she lacks the nuance that comes through touch, she tends to give her passengers whiplash. She has gotten into a couple accidents when her foot has slipped without her noticing. Her case shows us that the skin sensation that may at first seem superficial also contributes to her perception of her body that we'd tend to think of as proprioception.

Touch is crucial to our capacity for basic movement, but that's probably not what most of us think about when we imagine what it's like to lose the sense of touch. What's more likely to come to my mind is what it would be like to have none of the more delightful sensations of our lives—the caress of a loved one, the feeling of crawling into clean sheets, smooth sand under our feet, and the rumbling of a bike on a gravelly road. It's those tickles and nudges from the universe that make us feel alive. Touch is the most emotionally affecting of our senses, so it's curious to me how little Waterman brings up these textures he's lost.

Because of the extreme nature of his disability, his view of touch is mostly utilitarian. He can't answer philosophical questions about it, like how he understands where his own body ends and the world around him begins. He can't speak to the way he experiences emotions when he can't feel them rising up in his skin. He doesn't think much about the scenes he sees in movies of people finally touching someone they love. It's been such a long time since Waterman has had his haptic faculties that he can't remember these subtle sensations or get upset that they're missing. He's most aware of what his disability prevents him from doing.

"I have lost the fabulous sensation of subtle touch that you have, the roughness of wood, the softness of skin," he says. "I mourn that loss. It's gone. . . . That doesn't mean that if I were outside, I wouldn't like to pick up a leaf and explore it with my hands. Sometimes I remind myself what the cat feels like by brushing its fur against my lips, where I do have touch. Feeling its wiry coat is fabulous. It's what makes it what it is. But I think of fine touch as a luxury, and I can cope without it."

The anatomy behind this softer side of touch was first studied in the 1960s. Two Swedish researchers, Åke Vallbo and Karl-Erik Hagbarth, found that the nerves responsible for touch activate the emotional centers of the brain in ways that are unique from the other senses. They were tooling around in their lab with a process called microneurography that allowed them to visualize electrical impulses from peripheral nerves when they were activated. One of them had stuck his own nerve with a needle that was connected to an amplifier that recorded the electric signal that was generated. He was taken aback when he saw two of these signals radiate up after his skin was slowly stroked.[5] It was already common knowledge that pain produced two distinct sensations—the initial sharp prick followed by a protective diffuse wave because of the combined activity of fast- and slow-acting fibers. But he wondered why that was also the case for gentle touch.

Years afterward, Vallbo continued to study this reaction with other collaborators, including Håkan Olausson, who was also at the University of Gothenburg in Sweden. They noticed some interesting quirks about the second ripple of neural activity. It was strongest when very little pressure was applied and when it moved across the skin slowly, at a speed of about five centimeters per hour. It also happened only when hairy skin, not the glabrous, was stimulated. This was the exact motion of a gentle, relaxed caress. Perhaps this response exists to reward social connectedness, they guessed. Humans have evolved to coexist with each other and rely on each other for support, so this impulse for physical closeness could have been evolutionarily helpful. Touching those we love, after all, is our primary way of calming ourselves down and assuring our safety and security. The scientists called the fibers that produced this effect C-tactile afferents. While our brains attach emotional reactions to all of our sensations, these findings showed that touch is different because emotion is encoded from the very beginning.

The big unknown is how exactly this activation feels. The problem is that the activation of our fast-acting fibers, which tell us about the basic facts of whatever we're touching, cover up our awareness of the afferents flaring up. When someone we love strokes our arm, we might experience some pleasure. But that emotion isn't distinct from the physical feeling of being touched. For years it was impossible to study the impression the afferents produce on their own because they can't be activated independently. That changed when cases like Waterman's first became publicized.

In 2002 Waterman went to Sweden to meet Olausson and his team.[6] Because Waterman has no fine-touch sensation but can still feel pain through his intact C fibers, the researchers thought he would be a perfect case to test how they function. When Waterman wasn't looking, members of the team swept his forearm with a small artist's brush. Then they asked Waterman if he could feel it. Waterman said over and over that he couldn't. But when he was forced to answer if he was he being stroked or not, he was almost always correct. When he was placed in a brain scanner, the researchers confirmed there was indeed increased activity in parts of the brain known to receive messages of pleasure from touch. They deduced it was possible that, even though he didn't actually feel touch, somewhere beneath his perception the C-tactile afferents did put him in a blissful state.

The researchers began to believe that the second wave of physical sensation from gentle touch produces not a visceral feeling but an almost-sensation of emotional closeness. But Waterman remains skeptical. He says it doesn't really feel like much of anything, and because of that, he's pretty sure it doesn't motivate his behaviors in life. Then again, it's possible they do and he just doesn't realize it or has worked so hard on rationally thinking about how to replace his sense of touch that his mind can't make space for its emotional current.

"If all I'm doing is attending to subtlety, then I could probably get it," he says, skeptically. "But if it was that subtle in everyday life, I could not feel it. If I don't feel it, then how is it going to motivate me?"

There is a group of families in Sweden that has the inverse condition of Waterman's.[7] They can feel fine touch, but they don't have functional C fibers that are responsible for the second swell of pain and pleasure due to an extremely rare inherited mutation. The loss of pain sensation is the most debilitating part of their condition. Because it doesn't hurt them to break a bone or burn themselves on a stove, they don't know when damage is done to their bodies. This can be incredibly dangerous. But studies show that they also experience less pleasure than control subjects when their arms are caressed. As a result, they might be less likely than the average person to reach out to their loved ones in an affectionate way.

India Morrison, a principal investigator in the Embodied Brain Lab at Linköping University in Sweden who has been studying these families, is hesitant to make any clear links to their behavior from their biology, which would be something like assuming, based on a single example, that all women are bad at driving. She finds that they are generally within a normal range when it comes to their social interactions, especially within the context of the northern part of Sweden where the population as a whole is known for their chilly demeanor. What might differ is their reasoning for making these gestures. They might be trying to follow social graces rather than seeking warmth or closeness.[8]

The feeling created by the C-tactile afferents remains elusive. But however significant they are to our behaviors, it is beautiful to consider this in-built system that motivates some of our most visceral reactions and affectionate relationships. Touch is by nature deeply emotional, which is why we may think less about the practical losses someone like Waterman experiences and more about what it would be like to go without some of this magic.

In the late afternoon, as the light outside dims, I walk with the Watermans up a hill to their small turkey farm. They sell rare breeds that haven't been homogenized by commercial farming to customers looking to raise their own. Less often, they sell their eggs and meat. Handling the turkeys takes Waterman back to the old days at the butcher shop. Brenda goes around to each of the pens to feed the birds and passes the eggs to him as he follows behind in his cart. She makes sure that she allows him to do any of the work that he can, even if it takes him more time.

He cups each egg in the palm of one hand while he records their number and date with the other.

"For many years I couldn't pick up an egg," Waterman says. "Now, I only break two or three a year. I have to be careful, though. I can't take liberties. It's something I have to attend to quite a lot."

Waterman tells me he met his wife through work. At first their messages were professional, then they got friendlier. By the time they met in person, she didn't know about his handicap although she could see he had some kind of disability. She wasn't comfortable asking back then, and he didn't think it was serious enough to tell her. He said vaguely that he has a rare neuropathy. It never bothered her, except when she always found him asking her to get out of the car and fill up gas. When he explained that it was because of his condition, she understood.

When their relationship got more serious, he suggested she watch a television documentary about him so she could have the answers she needed. But by that point, she felt so happy in the relationship that she didn't want any preconceived notions in her head that might worry her unnecessarily. Instead of watching the program, she decided to simply observe his behavior on her own. Within six months she was confronted with just how serious his disability was. Waterman developed an ear infection, and the pain kept him up for three straight nights. Without rest, he was completely incapacitated.

"That's when I saw how much mental energy it took for Ian," Brenda says. "He worked so hard at hiding it that I didn't realize the extent of his disability."

He tried to grab a cup from his bedside table, but he kept missing it. When he stood up, he staggered. When he wasn't in top form, tasks he made look easy in most of his daily life were impossible. The illness passed, and he and Brenda were able to return to their normal activities, although the episode did make her understand his condition better. It made her realize, for example, why he was so addicted to chocolate. He used so much energy that he needed the constant rush of sugar. As they've gotten older and his age has drained him of some of his ability to focus, leading to more small errors in his movements, she encouraged him to finally start using a wheelchair full time. He was reluctant at first to give up the autonomy that he worked so hard for, but now he admits it has made his life easier.

Finally, I ask the Watermans the question that has lingered in my mind ever since I booked my trip to England: What is sex like? I can tell they've been asked about it many times before. Even still, Waterman is sheepish. His sex organs work normally since his motor function remains unaffected. But like anything else physical, sex requires extreme planning and concentration. I imagine him thinking about every step, unable to surrender to the passion of the moment. He admits there were some awkward discussions toward the beginning about the logistics.

"To undo a bra strap I couldn't see is impossible," he says. "So she either has to find a front-loader or I had to get her to do it herself. And I have to keep the lights on. Brenda thought that was a bit strange in the beginning, but I can't do anything in the dark. I've got to be able to see myself."

But what does he feel? The pleasure of sexual touch remains mostly mysterious. The skin of our genitals isn't well-equipped, like our fingertips, to detect edges or textures. Instead, it's covered in what's called free nerve endings, which transmit heat, cold, pain, and inflammation. It also has special types of nerve endings called end bulbs, which appear prominently in the sensitive skin of our lips and our nipples.[9] Waterman retains the former but doesn't have the latter. So for him, sex gives him some combination of heat and faint pain. It's still unclear which type of nerve activation is most important to the act of sex. Because he has almost no sensation, any change at all that's not painful is pleasurable to him. But Waterman also insists that it's not just about the physicality of the act.

"It's that somebody would trust you enough to engage with you like that," he says. "That means the world to me. You get trapped into thinking you can't be intimate because there's no touch, but that's rubbish. It's not just about that."

Brenda concurs.

"I still stroke his arm," she says. "In the early days, I'd want to hold his hand. But if I did, I could pull him out of balance. Once I knew the logistical reasons why we couldn't touch like I was used to, we just sort of readjusted. If he's safe in a chair, then we hold hands. But I would say that for both of us touch is more of the visual reaction of the other person. It's that he can see my warmth and my emotion, and I can see it in him too."

In his relationship, just as in the rest of his life, Waterman has adjusted to his lack of touch by using vision instead, and his wife has too in small ways. But for the rest of us, the emotions elicited by touch are some of the sense's most salient qualities. Only part of this is because of the nerves covering our bodies. These touchy-feely aspects are even more pronounced in our minds.

3

Mushy

When Sensation Crosses Into Emotion

It was during a conversation with some of her fellow resident assistants while in college at the University of California, San Diego, that Alicia Elba Williams realized there might be something different about her sense of touch. They were casually talking about synesthesia, a cross-wiring between the senses that causes some people to make arbitrary connections, such as between numbers and colors. The number eight might always appear to them in the color green. Or, in the case of auditory-tactile synesthesia, hearing the blare of a trumpet could provoke a total body itch. Some in the group were just interested in neuroscience. Others were pretty sure they had synesthesia.

Isn't that so weird?

Yeah totally . . .

But then, doesn't everyone sort of have that, to some extent?

Williams doesn't remember the exact words that were exchanged, only that the banter went down the same track as so many other discussions about the condition. But she has a much sharper memory of what happened next. Williams, a quiet, sensitive linguistics student, asked if anyone had heard of calendar synesthesia, which she has. It makes her think of time as existing on a spatial map of physical locations from her life. November is one place; December is another. She can mark important events that have happened to her along this neighborhood of her mind. Her synesthesia helps her with remembering important dates.

Oh yeah, I've heard of that, some of them said. A few thought they had it too.

While she had everyone's attention, she asked if any of them had another type of synesthesia, between touch and emotion, "like, when you touch a button or something and that makes you feel a really strong emotion." Williams described how certain surfaces could make her have highly specific reactions, from blissful to troubled. Sometimes they would bring back a memory. She suspected she would get some nods of understanding. Instead the room went silent as everyone considered the question.

"Nobody else has that?" Williams asked. They shook their heads.

As the group broke up at the end of the night, a friend of her named David Brang pulled her aside to chat more. He worked part time in the lab of V. S. Ramachandran, a prominent behavioral neurologist who directs the school's Center for Brain and Cognition, and had been looking for his own project to pursue with Ramachandran's help. Brang was interested in studying the form of synesthesia she was describing, as he was already developing an expertise in the condition, and asked her if she would like to come to the lab. She immediately agreed, curious to learn more about how her mind worked.

At her first session, Brang and Ramachandran asked her to handle textures with the pads of her fingers and describe how they made her feel.[1] She told them that corduroy made her confused, as if she were walking into a room without knowing what she was there for. Leather made her displeased, like she was receiving criticism. Moist soil was associated with contentment. A certain grain of sandpaper gave her the same feeling she might have if she were telling a white lie. Tylenol gel caps evoked jealousy, and warm water led to disorientation, like waking up after dozing off somewhere unfamiliar. Bok choy was disgusting, which was interesting because she had no problem eating it, just touching it with her fingers. Denim was the worst texture for her. It caused self-loathing. She told the researchers that she remembered getting into rows with her parents when she was in elementary school because she refused to wear jeans to school, even on cold days.

The researchers wanted to make sure her responses were consistent, so they had her come in for a few more lab visits months apart over the course of a year to ask her the same questions. The sandpaper that she'd originally said made her feel like she was telling a white lie, upon second consideration, made her feel "guilty, but not a bad guilt, like you know you had to do something wrong to get something better in the long run." The first time, a tennis ball reminded her of having fun with her dad. The second time, it brought up a more specific memory of listening to him play the guitar. Brang and Ramachandran believed her answers only confirmed how real and visceral her emotions were in response to touch. She was having the same feelings, and the fact that she was expressing them differently made it clear the associations she was making weren't just linguistic in nature. They were further convinced by examinations of her facial expressions and heart rate and a skin conductance test, which confirmed she was in fact experiencing what she said.

At the time of the study, there was no name for the type of sensory blending that she was describing. When Brang and

Ramachandran's work was published in 2008 in the journal *Neurocase*, Williams, who was referred to just by her initials, AW, and another participant, someone else Brang had met at the university, became the first-ever diagnoses of touch–emotion synesthesia. About a decade later, they remain the only published examples of this condition. There have been no other papers on this unique form of synesthesia. Most people, including experts who study synesthesia, don't even know it exists.

"I think it was a real learning experience for me to have to describe my emotions at the lab," Williams recalls of her experience on the study as we sit together years later at a cafe at Union Station in Washington, D.C., over tall steins of beer. "It's very rare in your life that you have someone telling you to describe what kind of anger or what kind of grief or what kind of guilt you're having and having to walk through that really carefully and very specifically."

It is after her workday at a government agency, and she is killing some time before taking the train back to her home in Baltimore. She is wearing a fuzzy red sweater and a black-and-white wool skirt, an outfit that is ladylike and vintage-inspired. She has a wistful, unhurried manner that seems to run counter to living in a city and to modern life in general, maybe because she is so used to stopping and exploring how she feels at any given moment. If anyone made a movie of her life, she'd be played by Zooey Deschanel.

Williams takes me back to her earliest memories of her unique synesthesia. It's hazy, she says. As a child, she had an ever-present feeling of being overwhelmed. Everything she brushed up against had so much emotional resonance for her that she didn't know what she should be paying attention to or how to manage her emotions. She often got swept out of the moment to explore these feelings, which made her disoriented and confused. Her skin operated as an emotional weathervane.

"Everything was sensory overload," she says. "I look back, and I'm like of course I was having nightmares because everything was getting triggered."

Her parents remember her describing textures and surfaces in ways that were so much deeper than they felt physically. For instance, she would touch water and say it was like she was lost. To her brother, water was just water, and it was wet. Williams got caught up in details that others wouldn't notice. There was a time before she was in school when she got startled on her way up the stairs because she felt like two steps were different sizes. When her mother took a closer look, she realized Williams was right, but the contrast was so slight that no one else had cared.

Williams's parents never took her to a specialist or discussed her temperament with her pediatrician. They just thought she had a brooding personality and helped her to calm down when difficult emotions swelled up. Williams's grandfather, her mother's father, was a number–color synesthete, meaning that numbers appeared to him in a consistent hue, so they knew that synesthesia runs in the family. But her mother, Daphne, didn't think of what her daughter described as a form of synesthesia, probably in part because that particular type hadn't been studied yet. Daphne also didn't think her daughter's thought processes were all that odd. Daphne, too, makes associations between emotions and things she felt, just different ones.

There are two main theories for why synesthesia happens. One is that we are all born with vision and sound and touch and the other senses blended together in a single stew. It's only in our first years of life that they differentiate from each other through a process called synaptic pruning. The connections between the senses that are most useful to our survival get preserved while those that the brain finds less germane get snipped away. For synesthetes, this process may not occur as completely as it does for the rest of us. The other theory is that synesthetes are simply born with more crosstalk between the senses than the average person, so even if they go through some pruning, more connections remain as they proceed through life.

The most obvious example of the connections we all make between touch and emotions is that physical injury causes us anguish. But there are other, subtler connections as well. We have an automatic aversion to abrasiveness. Rubbing against something rough might not cause immediate harm, but continuous exposure could cause painful damage to the skin. It could be that, recognizing this possibility, the human brain evolved to make grainy textures unappealing to encourage us to avoid them. In the same way, it might have forged connections to make us crave softness or smoothness because these textures remind us of human skin, and closeness with others is a way to improve our chances of survival.

What makes Williams different from the rest of us is that she has reactions to stimuli that aren't immediately recognizable as important for avoiding harm or motivating pleasure. The connections she makes seem fairly random, and the emotional qualities of touch are far more magnified in her brain than are the physical qualities. Williams's synesthesia might sound idiosyncratic, but it could highlight behaviors that are quite common in us all. We probably do the opposite of what she does. We focus so much on the physical that we ignore how much we're affected by the emotional attributes of everything we touch.

Williams often gets confused between the emotions that come from what she touches and those that come from real problems she's facing in life. Sometimes she's not sure whether she's feeling discomfort because she brushed against someone's jean jacket or because she had an awkward exchange with a friend. The emotions of both are just as real and important to her. The same is also true for positive feelings. She can easily overcome a bad mood because touching something pleasant can make her forget all about her more difficult problems. Putting on a silk dress or lying in a sleeping bag is an immediate pick-me-up. We all try to distract ourselves when we're unhappy, with a long nap or a tub of Häagen-Dazs. Such things can help temporarily, but we also know that underneath that lazy or fatty goodness remains real distress that we eventually have to face. But for Williams, her temporary distraction can actually supplant her true feelings. I'm a little jealous when I hear this because I'm her exact opposite. When something is picking at my mind, I have a rigid inability to move on, no matter how marvelous my surroundings are.

"When I wear Dupioni silk, which is a really special kind of Italian silk, it makes me want to dance, like being on the dance floor and sort of everything's gone away, and it's just pure joy," she says, her smile lines carved deep like a marionette's. "Whatever bad happened that day, I can still be like this is going to be the best night ever."

But she tells me that her emotional plasticity comes with consequences. Our moods perform important functions in our lives, helping us manage our relationships and make tough decisions. She has to make sure she's carefully parsing out which ones she needs to listen to and which are okay to ignore. When she was younger, she used touch to calm herself every time she got worked up. It's all too easy for her to excuse away a pressing life issue by telling herself she must have touched the wrong thing. But over time she's trained herself to think carefully about what she's upset about. When she recognizes that her emotions are telling her something important, she makes herself feel them. But this is an act of serious discipline.

"If I'm dealing with feeling guilty, it can be easier to just blame it on touching the wrong thing and say 'Okay, let's ignore that. Let's touch something else,'" she says. "It's hard sometimes to really pull aside and get to the core of the emotion. So that's been a real learning experience."

Emotion plays a major role in what sticks in our brains. We remember events that get us riled up much more than mundane ones. Because for Williams tactile moments are imbued with so much feeling, she can recall them distinctly years later. This may have been why she developed an association between the texture of a tennis ball and being with her father. It could also be why she so vividly remembers the weight of her grandparents' silverware or the shape of a button on one of her grandfather's coats. Now that he has passed away, touching a peacoat button will make her instantly nostalgic for him. Sometimes her brother asks her to recount these memories for him, amazed at what she's able to recall. She considers this mental capacity for tactile details to be a superpower.

As we're talking, I notice her face change all of a sudden. She says the texture of the tablecloth is reminding her of baking cookies with her mother when she was a little girl, and she's drifting off into the past and getting nostalgic. I notice her hands playing with the copper alloy pendant of a wing nut and screw on her necklace. She says she always wears jewelry or a watch because running her hands on the metal is calming. She does it without noticing it. It's like a rest button that brings her back to the present.

"Metal is one of the few things that doesn't make me feel anything," she says. "It's calming and cool. I've always used metal for sort of re-centering and refocusing and getting rid of emotions. I touch it, and it's just done."

I realize in that moment just how constantly Williams has to manage her synesthesia, which is really just a hyperactivity of the same brain mechanisms we all use when we interpret what we touch. There is something that seems familiar to us about Williams's condition because it's an exaggerated form of how we all link together touch and emotions. Long before we had studies on synesthesia, we've known that touch is a sense uniquely connected to our psychology. One of the earliest examples of literature exploring this innate awareness comes from Aristotle.

De Anima, Aristotle's treatise on the nature of the soul, highlights the way we yoke them together. He thought that by letting us know where we end and the world outside of us begins, touch helps us to develop our self-awareness, and it's this knowledge of our existence that makes us want to impose our will on our surroundings.[2] It makes us feel the hunger that drives us to go hunt and the curiosity that pushes us to explore. It's the reason we seek shelter and develop bonds with others. Without the feelings that come from touch, we'd be passive beings, waiting around for nature to meet our needs. To Aristotle, this was the distinguishing factor that kept us in a class separate from plants.

Of all animals, he was sure that humans have the keenest sense of touch, which gives rise to our superior intelligence. We have no thick skin, scales, or fur to protect us. We are naked and exposed, which makes us likely to get injured. We get hot and cold too easily. We experience the extremes of pain and pleasure. Our lack of defenses forces us to be more careful and resourceful than other animals. He thought these physical qualities are what make us emotionally sensitive, which lets us connect deeply with the suffering of others. For him, the knowledge of our own existence that touch provides has a direct relationship to our capacity for mental self-reflection, our most important and human of qualities.

He writes:

> With respect to the other senses, man is far inferior to the other animals; but with respect to the sense of touch he excels by far in discrimination over the other animals. This is why man is the most intelligent of animals. A sign of this is the fact that, even within the human race, it is by virtue of this sense organ and of no other that some are well-gifted or poorly gifted by nature; for those with hard flesh are poorly gifted for thought, while those with soft flesh are well gifted.[3]

In the days before surgery to examine the insides of the human body, Aristotle was making an instinctive statement about touch, in direct opposition to his mentor Plato, who thought of it as a primitive sense. Aristotle thought of our capacity to touch as the source of our ability to feel, and to intuit the same in others, our greatest intellectual gift. This was a revolutionary idea, presenting the body and its emotions as superior to our logical decision-making, represented by the mind. In the end, he didn't win this debate in the culture at large. Thinking is still considered our highest function, and we continue to see as distinct the feelings of the body and our emotional feelings. But even still, Aristotle's reasoning has had resonance.

Touch and our emotions both have to do with changes in our bodies, including our skin. All of the senses can elicit our emotions to some extent. A delicious meal or a beautiful aria can bring us to tears. It can be argued that smell has a particularly intense relationship to our gut reactions. It causes extremes of pleasure and disgust. But although these feelings are immediate and instinctive and perhaps even more potent than touch, they fall within a limited range. With its complex cluster of different types of sensations and reactions to those sensations, touch represents a larger palette. It's the source of our worst pains and greatest joys.

This makes me consider whether sensation can similarly act as a window into an animal's psyche. This is the thought swirling in my mind as I meet Duncan Leitch, a postdoctoral scholar in neurobiology who studies animal touch at the University of California, San Francisco. We're just above the habitat of an albino alligator named Claude at the California Academy of Sciences. Claude is in his indoor swamp near an overhanging bald cypress tree. He is pale and motionless, wearing a perpetual wide grin. He looks like a happy vacationer, his large body splayed out on top of a rock like he's sunning himself. His eyes are open so it seems he should take notice of the fish and turtles moving around in the water underneath. Instead he sits there, unimpressed.

Leitch is an expert on the specialized touch sensors of crocodiles and alligators and has spent many hours trying to enter these animals' bodies and minds.[4] Touch is a sense that's invisible to the eye, so he often has to intuit how it works through the animal's behavior. He empathizes with his research subjects. It's something he's done since he was a kid. Growing up in Memphis, he would play around with tadpoles and turtles and other creatures he came across at the park and try to see how they solved the same basic problems as us—eating and finding shelter—just in dramatically different ways. They fascinated him in a way that cuddlier animals like cats and dogs didn't because their behavior seems so foreign.

He points out the raised dots along the middle of Claude's face, a feature I'm familiar with from many a designer handbag, imitation or not. Scientists once hypothesized that these bony nodules could detect electrical fields or maybe salinity. But in recent years Leitch and others have found that they are packed with special nerve endings for the mechanical, chemical, and thermal aspects of touch, which is noteworthy because humans have separate sensors for each of these. There's an obvious need in alligators for a tough exterior that can survive injury. These sensors are little chinks in the armor that provide them with exquisite sensitivity.

"They're very, very sensitive," Leitch says, comparing their mid-face to a more reactive version of our fingertips. "It might be akin to if we had super good vision. It's not that if I see something that's bright, it's so bright it's painful to me. It's just that I'd feel like the awareness was larger."

The reason we so often see alligators with their head half in the water is so they can use these bumps to feel every minute ripple to identify and capture prey. These features are also important to their social lives. During courtship, alligators like to nudge each other or touch their snouts, kind of like human beings holding hands. Even though reptiles aren't known for their maternal instincts, alligators are an exception. When their babies are hatched, the mothers carry them around in their mouths to protect them, using their acute touch to keep them away from their sharp teeth.

During the breeding season a male alligator will make vigorous vocalizations under the water at frequencies we can't hear but that can be felt above ground. The female uses the vibrations to tell whether he is worth her time. The stronger they are, the better chance they have of getting together. When we think about it, this is not too unlike how human speech works. When communication has to occur through a body of water or underground, the feel of the vibrations carries much better than sound waves would.

As Leitch waxes on about an alligator's sensitivity, I notice something happening. My entire orientation toward Claude changes. Knowing how he feels in his skin seems like a window into what it's like to be him. It's more evocative than knowing how he sees or hears or tastes. It gives me a vague outline of his emotional life. Claude is still in his tableaux position, but in my mind he's no longer a boring, sedentary monster. I see a creature so soft inside that he needs a hard exterior to get by. He's practically cuddly. Claude starts stirring. He swings his head. His white skin gleams under the bright lights. He dives off his perch into the water in a moment so magnificent it appears in slow motion even as it's happening. A baby's breath bouquet of air bubbles flies up around him. When I ask Leitch about the touch–emotion connection, he says, "It might be something in the extremes of sensation with touch because things like pain are so visceral, and there are pleasant components as well. And there's also a range to what we feel. There's a nuance."

Leitch shows me around the rest of the animal exhibits, and each time I have the same response. He points out the Mexican red-knee tarantula, furry and black with bright bands of orange at its joints, which are highly attuned to movement. They help it feel for what's happening underground, including the stirrings of prey and the messages sent by other spiders. I picture the harried mind of this lively little animal, how agitated it must be, constantly having its calm interrupted by elbows and jabs from the earth, its stress levels steadily ratcheting up. We approach a tank containing an elephantnose fish, which has skin covering its eyes, so it relies almost entirely on touch. It is an active electrolocator, a fish that generates an electrical field as it swims about; it uses this field to sense and locate objects in the environment.

Leitch pulls up on his phone a picture of another animal he has studied, the star-nosed mole, which is considered to be the most sensitive mammal of all. It is about the size of a hamster with the body shape of a seal and the fat paws and long claws of a sloth. Although most of its body is no more sensitive than ours, on its snout is a special touch organ that looks like the delicate underside of two side-by-side starfish. The rayed appendage is about the size of a pinky nail and is innervated ten times more than an entire human hand. As a microscope is to vision, a star-nosed mole's organ is to touch. It can sense even the slightest movement underground. The star organ likely evolved to help it sense the textures and vibrations of tiny insects that its competitors, such as other moles, voles, and snakes, can't find. It's the freegan of the animals, making do with the leftovers that nobody else wants.

Although, as a scientist, Leitch tries to picture what it's like to be one of the animals he studies, he stops himself from anthropomorphizing. That's not because he thinks the animals don't have rich inner lives; it's just that he couldn't know what that would be. But clearly that doesn't stop me. I know that I'm projecting when I try to imagine these animals' minds and that I can never know the subjective experience of another. But still, the way I think about them changes completely. I can't help but connect to their vulnerability. At first, it might seem like a stretch to say that our sense of touch is at the root of such complex human traits as desire and empathy. But more and more, it doesn't feel so far off.

Touch serves a basic purpose in all animals; it tells us what we're up against. But there are a variety of ways it does this. There are animals with dull awareness and extreme sensitivity. Some have uniform feeling throughout their bodies, and for others it's localized. Some use it to detect what's near while others use it to know what's far away. There are animals with the ability to sense salt and electricity, and these count as part of the sense of touch too. Some even sense their surroundings by generating their own electricity. There are as many senses of touch as there are varieties of landscapes and body types in nature because how it works in each species depends on a variety of factors including where it lives, whether it's strong or weak, and how it manages its relationships.

For years, we've believed that our suffering is the most profound, that our appreciation of pleasure is deeper and more perceptive than that of any other animal. It's a self-serving idea, a convenient myth we've concocted about where our rightful place is, and it's something we're in the process of rethinking today, as a plethora of studies show that animals have a greater capacity for feeling than we've previously acknowledged. Part of the reason we've denied these feelings is because if we didn't, we would have to change our behavior toward them. It's telling that when we recognize that another animal feels, we have a responsibility toward it that is otherwise easier to ignore. Of course, that tells us more about our own sensory imagination, and how we've applied it in the past, than it does about theirs.

"It's a pleasant idea that animals can't feel pain to the extent that we can or they don't have emotions the way we do," Leitch says. "I guess it makes it easier to treat them in certain ways."

The way that touch and emotion cross over into each other affects our larger beliefs about the world. While this blending between the physical and emotional is something we've perceived in theory, until recently we didn't know why it happens. Now a growing body of work by neuroscientists is suggesting their connection could have been evolutionarily helpful. Before going into more specific examples of this work, it's important to note that priming studies, which aim to point out that small cues from our surroundings exert a large influence on our biases and actions, have come under some criticism in recent years because often their results can't be reproduced. Some of these examples could simply be cases of scientific overreach based on small sample populations and may be inconclusive. Taken as a whole, though, they do tell an interesting story that add another layer to how we experience touch.

Naomi Eisenberger, a psychologist and director of the Social and Affective Neuroscience Laboratory at the University of California, Los Angeles, first noticed the connection between touch and emotion networks in the human brain by accident, when she was viewing functional MRI images of research subjects who experienced social exclusion during a video game. She was sitting next to a research colleague who was at the same time looking at images of patients experiencing suffering from irritable bowel syndrome. Both showed blood-flow patterns that were remarkably similar.

Eisenberger started to believe that there might be neural correlations between emotional and physical discomfort; there's a reason we use the word pain to describe both. Her theory was confirmed when she conducted a study that found over-the-counter painkillers helped reduce both physical pain and social pain in research subjects.[5] When people were playing what was essentially a virtual game of catch and were left out, taking a pill made them report feeling less hurt by their exclusion. Her work provided more evidence that rejection can cause damage similar to physical injury, even if we treat them differently. This particular study did not account for the placebo effect. She started to brainstorm other ways that psychological experiences might match somatosensory ones.

In her next study she looked at the connection between physical and emotional warmth. Again her team placed subjects in the fMRI machine, which recorded their brain activity as they read neutral and loving messages from friends and family while they were holding either a warm pack or a ball. Participants reported feeling physically warmer when they read the positive messages from loved ones. Holding a heated pack made them feel more connected to their letter writers, even when their messages were neutral.[6]

After that she considered sensitivity. Her team started to look at whether people's resilience in response to social pain corresponded with their ability to tolerate physical pain. They exposed people to a thermal stimulus and varied the temperature in ways that some could perceive as painful. They compared the results against being left out of the video game. It turned out that people who were sensitive were that way all around, whether physically or emotionally.

Eisenberger began to believe that these connections must have been evolutionarily advantageous. It could have been that the pain of social attachment had piggybacked onto signals we already had in place for physical pain, which is useful given how important social connection is for human survival. The anguish of separation, exclusion, and loneliness are meant to provide a motivation to reconnect socially, especially in babies who remain reliant on their parents in their early years.

Eisenberger's studies deal with sensations, such as some types of deeper physical pain, that not every scientist would consider touch related. But other studies have found even more surprising examples of sensory and emotional blending that are purely tactile. People are more likely to rate a job applicant highly if their resume is attached to a heavy clipboard instead of a light one. In other words, this candidate carries more metaphorical weight or gravitas.[7] A group putting together a puzzle with rough pieces will describe the process as more adversarial than one using smooth pieces. People judge a boss as being stricter when they're told a story about his interaction with an employee while they're holding a hard block rather than a soft blanket. When we're meeting someone while holding a hot mug, we're more likely to describe them as having a warm personality.[8]

Many of our emotional responses to touch are automatic. Silk pajamas are usually going to cause us joy and sandpaper is not, so evolution could explain some more universal touch–emotion connections. How extreme these feelings are, however, depends on our individual bodies. Being born with slightly different tactile circuits, it turns out, can change our personalities and how we connect with each other. People whose skin is more sensitive to small changes in the environment tend to have more reactive temperaments. One way that researchers study personality in labs is through skin conductance tests that show how much people heat up and sweat in response to various stimuli, such as music or emotions. Their "cool" way of acting could be directly related to their physicality.[9]

Some people are born with a condition called mirror–touch synesthesia, which makes them see someone receiving a hug or stubbing their toe and have those same sensations themselves. This extremely rare condition, which affects about 2 percent of the population, is believed to be caused by overactivity in the mirror system, a set of neurons that react to others' actions as if they are our own. Having the condition is said to cause heightened empathy because it means being able to take on another person's feelings.[10]

According to Brang and Ramachandran, the number and intensity of our touch–emotion connections could underlie a talent for metaphor, the lynchpin of artistic ability. There are plenty of everyday figures of speech that seem arbitrary when we think about them, such as a *loud* tie or *sharp* cheddar. There's no commonsensical way to explain these links made between different perceptual realms in the English language, and yet they feel immediately obvious and right. The reason could lie within our brains. In synesthetes, this effect is likely magnified. Their unique wiring could make them specially equipped for linking different perceptual realms in creative and unexpected ways. According to one study, synesthesia is significantly more common in creative people, such as artists and writers, than in the general population. The colorful abstract artist Wassily Kandinsky explained that opera gives him visions of "wild, almost crazy lines." *Lolita* author Vladimir Nabokov saw the letter N as a "greyish-yellowing oatmeal color."[11]

Only part of the reason we connect touch to our emotions is because of the architecture of our brains, according to scientists. It's also something we learn to do from a young age. As children, we may have felt our mothers getting hot and shaky when they were placed on hold during a phone call. Calmness in comparison felt cool and smooth. We learned to interpret these sensations of the body as emotions in both others and ourselves. Before we could label them as emotions, we knew them by touch. As Ashley Montagu, the late British anthropologist, writes in his book *Touching: The Human Significance of Skin*, "Although touch is not itself an emotion, its sensory elements induce those neural, glandular, muscular and mental changes which in combination we call an emotion."[12]

Many of these learned associations end up in our language. Montagu lists several of them in his book.

> We get into "touch" or "contact" with others. Some people have to be "handled" carefully ("with kid gloves"). Some are "thick-skinned," others are "thin-skinned," some get "under one's skin," while others remain only "skin-deep," and things are either "palpably" or "tangibly" so or not. Some people are "touchy," that is, over sensitive or easily "irritated." Others are "out of touch," or "have lost their grip." The "feel" of a thing is important to us in more ways than one; and "feeling" for another embodies much of the kind of experience which we have ourselves under gone through the skin. A deeply felt experience is "touching." Pleasure in a work of art gives some of us "goose pimples." We say of some people that they are "tactful," and of others that they are "tactless," that is, either having or not having that delicate sense of what is fitting and proper in dealing with others.[13]

Because touch was already part of our mapping system for emotions, as we had more tactile experiences, we continued to add new data points. As we got older, we may have felt our knees rubbing painfully against carpet or a bath that was not quite the right temperature. Later, we applied those same feelings to describe what we were facing in our lives. We might have started chafing against expectations or feeling lukewarm about going to school. Through the course of our lives, we've felt the sting of defeat, admired the gritty style of a musician, received bruising lessons, and felt pressured to make tough decisions. Some linguists say that our language itself, particularly our use of metaphor, illustrates the way the sensory experiences of the body lay the groundwork for our more abstract thoughts.[14]

"The very structure of reason itself comes from the details of our embodiment," write George Lakoff and Mark Johnson in *Metaphors We Live By*. "To understand reason we must understand the details of our visual system, our motor system, and the general mechanism of neural binding."[15] What they're saying is that there is nothing in the mind that didn't start first with the body.

In English, the senses each have their own metaphorical associations. Just as vision has become a stand-in for thought, taste is symbolic of the ability to value all kinds of pleasures, from art to music. Smelling is about guessing. "I smell something fishy" or "I smell fear" indicate a dog-like attunement to mood or the state around us. Listening is equivalent to obeying, which can be a positive association in a culture that values conformity. When someone's not ready for your advice, "They just can't hear you." Touch is all about our emotions. It has the longest entry in the *Oxford English Dictionary*, most of which has nothing to do with physical contact but instead with the feeling of being affected by, say, a person or melody.[16]

Touch–emotion synesthesia may seem surprising or strange at first, but it highlights a universal truth. Our sense of touch shapes our emotional and mental worlds in important ways. Without the impulses for self-protection and self-fulfillment that it gives us, we wouldn't have the same capacity for reasoning, kindness, or compassion that we do. We need that literal skin in the game. In turn, the tactile language that we use to describe our emotions changes the way we feel them too. We think of emotions as states that occur from within, but they're inextricably linked to what happens outside of us.

A few weeks after meeting with Williams, I call up her mother, Daphne, in Murphys, California, a small gold rush town in the

Sierra foothills where she raised her children and still lives. Daphne is a language teacher at the local high school. I'm curious about what it was like for her to learn about Alicia's experience with the synesthesia study. Williams's involvement in the study actually made her come to the conclusion that her mother also had it all along without realizing it. Interestingly, it turns out Daphne has a different take. She doesn't like the idea of having a psychological diagnosis.

"I just don't think I'm that different," she says firmly. She thinks that touch–emotion blending is completely natural, and that we all do it on a spectrum.

I learn that Daphne, like her daughter, has certain haptic preferences. Fabrics have attached meaning for her. Perhaps because she grew up knitting and sewing, she appreciates a hand-knit wool sweater, which makes her think about the origin of the yarn and the people who work it into something special.

"It's heartwarming. It's safe. It's like coming home," she says.

She doesn't like the feeling of fur.

"It's intrusive and unpredictable and dirty. I have dogs, but this is going to sound weird. It's not a comforting feeling at all."

Once she confused her dentist by describing a pain in her tooth as "velvety. It has grooves in it."

Daphne remembers her husband watching curiously once when they were out shopping for dinner plates. She quickly picked one up and then rejected it. He asked her how she was making decisions. She tried to break it down for him, saying that in some cases it was about the finish and in others it was about the way an edge felt.

"Oh, you like it because it has that design," he offered, glibly oversimplifying what she said.

"No, it's not the design; it's how it makes me feel," she told him. "This plate is more comforting. It makes me feel good."

I tell her how much she sounds like her daughter. Williams had told me an almost identical story about going shopping for mugs with a bemused friend. She, too, became exasperated at having to explain why one just seemed better than the other.

"Don't people always like what they like for reasons they can't explain?" Daphne insists.

She asks me what other people feel when they're picking up plates if they're not feeling emotions.

"Sometimes we just don't feel very much," I say, adding that I can really only speak for myself. "I touch a plate and I'm aware of nothing else besides the knowledge that I'm touching a plate. It's just, 'that's that.' If I were buying tableware, I'd probably get it online because how it feels doesn't matter all that much to me."

We stay on the phone in silence for a second before she declares again what she said at the beginning. "I'm not sure I'm really that different from anyone else."

Later, with a group of friends at dinner, I bring up how I shop for plates and cups, and I discover that my approach is not actually that common. Some friends are astonished that I buy them based on a picture without ever holding them in person. The heft and the feel of their finish matter a lot more to other people than they do to me. Suddenly I feel like the weird one. It's certainly possible that I underplay the emotional component to what I touch as much as I consider the synesthetes to have an outsized reaction to it. It's always there. It's just a matter of how much we let ourselves notice it.

I'm not a doctor. I can't say whether Daphne has the same synesthesia as her daughter. But if she does, it's worth considering which is a more useful way of thinking about it: Daphne's sense that it's somewhere along the spectrum of normal or her daughter's concept of it as a unique condition that needs to be managed. I feel myself slowly moving over to Daphne's perspective only because I think it's useful for someone like me to recognize all the subtle signals I'm constantly getting through my skin and my body that I might be ignoring.

Even though we're constantly mapping our emotional worlds onto our physical experiences, we aren't consciously aware of it. One reason may be that much of the brain's processing of tactile cues occurs in the background. It's doing this work while we're busy with other things. So even though we may ascribe our behaviors to logic and reason, we're more influenced by our surroundings, particularly what we touch, than we recognize. The texture of our clothes, the temperature of a room, and the dryness of the air all feed into how we feel at any moment. When we try to notice, by the minute or the hour or the day, how exactly our surroundings are affecting us, we can't.

4

Untethered

Will the Body Become Obsolete?

I'm confused when I find myself in a large parking lot looking at a shopping mall. Google Maps was supposed to direct me to the Simon Fraser University School of Interactive Arts and Technology, but I don't see any red-brick buildings or students lugging around backpacks. There's a bus station and folks with shopping bags. It's a typical British Columbia suburb. But my phone insists that it's right, so I listen to it and head toward the mall. Sure enough, before I hit the lines of stores selling cheap teen fashions and the food court, I look up and see the campus tucked into the corner of the complex in a series of office spaces.

"People are always surprised when they visit. The Europeans love it," says Diane Gromala, a digital artist and professor who greets me at the entrance. Gromala and her husband, a computer scientist named Christopher Shaw, run a lab here called Computational Technologies for Transforming Pain that designs virtual reality games that help people with chronic pain, a condition that occurs when any pain lasts for more than three months. Often it starts with an injury but then ends up persisting long after the body is done healing, becoming its own separate problem. The greater the emotional torment about the pain, the larger the pain looms, creating a never-ending feedback loop that can trigger the corrosion of the gray matter responsible for cognition and psychological issues, such as anxiety and depression.

Gromala had initially hoped to create programs for pain sufferers as a distraction mechanism. It was based on previous design work she'd done with the University of Washington virtual reality researcher Hunter Hoffman on SnowWorld, a game that helped burn victims distract themselves from the pain of having their bandages changed while they were in the hospital. She thought the benefit would be similar for pain patients. While they were occupied with something as novel as a virtual reality headset, their negative mental cycle could temporarily be put on pause, a crucial step on their path toward recovery. But as she started testing the program on subjects, she saw something even more captivating happening. These patients were actually dissociating from their aggrieved bodies and identifying with the avatar version of themselves that was healthy and powerful.[1]

Gromala's research is a thought experiment for me. A dystopian take on our virtual future comes from the movie *Ready Player One*: we'd get so caught up in the better-than-life fantasy world we see through our goggles that it would be the only place we'd ever care to be. There we wouldn't have to follow rules like gravity and the limits of our own abilities and the law. We could be anyone, anywhere—more attractive and fitter and happier. As we spent more of our time there, we would forget all about our real bodies and actual lives, and we'd let them both crumble apart. This story is not just about the future. It's about people's worries that technology is overtaking our present.

Convincing virtual reality has promised to be just a decade away ever since the 1980s, and it still doesn't seem we're that close.

But we are already slipping further into our digital lives and, according to many people, too far. We are surrounded by ever brighter lights and bolder advertising and larger screens that compete for our attention. This has very real consequences for how we feel. The more we use vision to guide us, they say, the less in tune we are with our bodies. Touch and vision have long been in metaphorical opposition with each other, which we're now learning is because of the way these two senses work together.

Gromala, a woman with bird-like features and a nest of dreadlocks atop her head, is gracious but reserved. Her shoulders are stiff, her facial expression is taut, and she carefully considers each word as she speaks. As I spend more time with her, I see that what I'm interpreting as a guarded personality is more representative of the constant pain she's in—the kind that means she's not sure if she's going to make it to work from day to day and that sometimes keeps her in bed for weeks. Her chronic pain is the reason she started the lab. She wanted to help people like her and to have a job where people understand her limitations.

She leads me to a classroom whose perimeter is lined with computers. One of her grad students helps strap me into a virtual reality headset so I can try out the game for myself. I have a view of a dark cartoon cave with rocky walls on the sides. When I turn my head, the scenery moves with me, which gives the impression that I'm surrounded by this animated environment, not standing amid a bunch of students working on their own projects. This is the first time I've tried out an Oculus device, so the sensation takes some getting used to. Cute little angry ghosts pop up in my field of vision.

Another of her students pipes up to tell me what I should be doing. Apparently if I turn my head, I can aim fireballs at them. Only when they're dead will I be allowed to move forward through the rest of the game. I feel self-conscious, knowing that everyone in the room probably has their eyes on me while I can't see them. I'm even more paranoid that one of them knows what's on my screen and can see how much I suck at the game, so I devote myself completely to the task. It takes a few seconds to get used to coordinating my own movements with the images I see, but soon it feels as natural as the illustrated backdrop does.

The shooting practice has allowed me to align myself with my avatar, and it becomes no more complicated to turn and strike a ghost than it is to tie my shoelaces. We're so connected, me and this representation of me, that at about a minute in, I move away when virtual objects come hurtling in my direction. I'm sure I look crazy as I duck and bob, but I don't really care about the world outside anymore. The cave's ceiling appears a little low, which makes me feel claustrophobic and nauseated, even though I know intellectually I could walk right through it if I wanted. I see a light at one end, and I follow it all the way out.

I enter a new setting, a snowy pathway lined by undulating trees with angry-looking branches that heave around menacingly. When I hurl at them accurately, they tuck their arms to their sides, which gives me enough room to slide past. "The trees are a visual allegory for angry neurons that I'm treating with the anti-seizure drug Gabapentin," Gromala tells me later. For people whose bodies are constantly flaring up with pain, I can imagine how empowering this task is. Being able to quiet down their pretend environment gives them a sense of agency that's missing in their lives.

"You really lose your proprioceptive sense," says Gromala, once I'm out of the headset. "When you're in this environment, it feels like the body map is malleable. You don't know where your edges are anymore."

The way chronic pain is treated has changed dramatically since Gromala started suffering from it back in 1984. It began as a throbbing sensation around her lower left abdomen and renal area. She went to see doctors about it, and initially they tried to treat her for other health problems, such as her endometriosis, calcifications in her uterus, and rheumatoid arthritis. But that didn't make the pain go away. At this time, pain was seen as just a consequence of injury and not a problem of its own, so the doctors were without options; they told her she should try to relax. Of course that was impossible. Gromala worried for years that the pain would get too unbearable and she wouldn't be able to work.

It wasn't until fourteen years later that doctors finally diagnosed her with chronic pain. By then the medical understanding of it had changed. Doctors were treating pain as its own disease and understood that its sufferers' thoughts about their pain play an immense role in how their brains let them experience it. A patient injured during a tennis match against a major opponent might will herself to feel better for an opportunity for another face-off. A patient who broke an arm in a car accident might blame himself for his shoddy driving and think of his pain as deserved punishment, causing it to continue. In addition to receiving painkillers, Gromala received recommendations for ways to calm her mind, such as yoga and meditation. She started to feel better as she became less panicked.

She believes virtual reality could be a potential new treatment for the social and psychological side of pain management. Gromala is an artist, not a scientist, so she tries her best to explain how exactly it works. Each of us has a body schema, or a mental map of where we stand in space, and it's highly malleable, she says. When we're using tools, they become part of the body schema. Same with our clothes; if we're in a puffy jacket, it changes the way we move within a space. This habit of our minds helps us adapt quickly to new tasks and environments. In virtual reality, the body map is stretched to its limit by allowing an avatar into its self-concept.[2]

The effect is created through giving the user a first-person view. But there are other ways to intensify it. The more an avatar matches our bodies, not only by following the position of our heads but also our eye and body movements and facial expression, the more we identify with it. The holy grail of virtual reality would be haptic effects that could make a user feel, say, the wind swishing past as they walk or the feeling of a ghost dissipating under their fingertips. If the feelings of the body are what chain us to the real world, then supplanting them with the atmosphere of the pretend one could be the key to make the illusion complete. We could leave ourselves behind completely.[3]

We don't yet know the consequences of a true out-of-body experience. We experience some discomfort when how we feel doesn't match what we see, which is why some virtual reality users report getting motion sickness while in the headsets. But even if there are ways to overcome this side effect, the next question is how long the illusion can sustain itself. It's possible that the brain can only believe what it's seeing for so long before the feelings of the body come back into awareness. On the other hand, maybe the effect could last forever. Then, would we have any need for our actual bodies anymore? Would they even continue to live on without the rest of us? And if our bodies don't matter, then where exactly can we locate our sense of self?

It may all sound hyperbolic, but these are issues experts will be busy with for years to come.

To figure out the answers to these complicated questions, we have to look again at how our senses function. In 1842 the German physiologist Johannes Müller proposed a bold idea: that our senses don't faithfully represent reality. According to his "law of specific nerve energies," what they tell us has less to do with the facts of the external world than with our nerves' own structure and behavior.[4] For example, the special qualities of the eye's nerves are what cause it to react to light the way it does. What we receive through our senses, then, is an impressionistic image of our surroundings. His law was a serious departure from the ancient and widely held notion that the nerves dutifully carry the properties of what they're sensing, and it was somewhat controversial because it hinted at just how fallible we are.

Müller's law applies to all of our senses. But many of his contemporaries in Germany and the United States independently came up with theories similar to his, based on experiments on human skin. For instance, Magnus Blix, a prominent Swedish physiologist, applied low-intensity electricity to his wrist and forearm with a small pin and noticed that he felt heat only in specific spots. If he moved the pin ever so slightly, the same stimulation made him feel a shot of cold. He realized that the skin isn't uniformly sensitive. It's made up of a patchwork of regions attuned to certain types of sensation. He and others mapped out zones with sensory specializations including detection of light touch and pressure and pain. They found that coldness, smoothness, bumpiness, and sharpness are products of the body's own machinery, and our experience of touch is dependent on the nervous structures that animate it.

The way that the senses coordinate with each other is also governed by our nervous system. Our sense of self is a figment created by connections between the brain, the spinal cord, and the sensory organs as well as the internal cues such as our breathing and heartbeat. Of the traditional five senses, touch and vision are the most important to self-awareness. Think about how we become conscious of our own existence as separate from our surroundings; it's when touch and vision are aligned. When we jump, our eyes observe the scenery shifting up and down at the same time that our bodies feel this movement.[5] Touch is really a two-part sensation. It tells us about our external surroundings as well as how they impact us. We feel the hard surface of the ground and the jostling of our own flesh. When we're sitting down, we see how we're positioned against the furniture and feel pressure in our skin. Touch and vision confirm each other's version of the truth.

The two senses' natural cooperation has made them a subject of interest for philosophers and scientists. A centuries-old thought experiment has tried to delve into just how deep their connection is.[6] In 1688 a Dublin politician and intellectual named William Molyneux, whose wife had gone blind shortly after they married, sent a letter to philosopher John Locke asking him whether a man who was born blind and who had learned to discern objects using touch would be able to recognize those same objects by looking at them if his sight were restored. In other words, without all that childhood experience that helps us integrate both of the senses, would we do so naturally?

If the answer were yes, that would mean that touch and vision must share some kind of internal logic that makes them relate to each other. If the answer were no, then it could be deduced that it's later on, through the process of learning, that our brains know to knit them together. Locke said the answer was no, a blind man with his sight restored wouldn't automatically be able to connect what he felt to what he saw. He wrote that touch, because of its association with bodily movement, naturally occurs in three dimensions whereas vision appears in two dimensions. Before anyone comes to see a sphere as an object with contours, it is just a "circle variously shadowed." It's only by seeing and touching at the same time that we learn how to interpret volume. "The Blind Man, at first sight, would not be able with certainty to say, which was the Globe, which the Cube, whilst he only saw them: though he could unerringly name them by his touch, and certainly distinguish them by the difference of their Figures felt."[7]

After he sent that answer, Molyneux's question spread through philosophical circles, and over years there have been many other takes. The Irish philosopher George Berkeley in the early 1700s agreed with Locke but for a different reason. Locke's assumption was that the senses infer some factual information about the objects they're experiencing; each of them just has a different way of coming at it. But Berkeley felt that the senses' connection to physical reality was irrelevant. What's present to the senses is all there is to speak of, and distance is not visible to the eye. He argued that vision and touch could be no more aligned than the flavor and shape of a banana.

Like Locke, Thomas Reid, a Scottish philosopher from the same period, said that the senses *do* represent reality. This meant that, like Locke and Berkeley, he landed on a negative answer to Molyneux's question because vision appears in two dimensions, not three. He added a caveat though. If the blind man in question happened to be a mathematician who already knew how to relate two-dimensional diagrams to objects with volume, then he might have the ability to see depth right away. It's just that it would be his knowledge of theory and not practice that would get him there.

A minority of philosophers have taken the positive position to Molyneux's question, arguing that a blind man would automatically connect what he saw to what he felt. Gottfried Leibniz, a German logician, mathematician, and natural philosopher, is one example. He thought the brain could reason pretty quickly, based on the way a cube appears and how it feels, that the two senses are observing shared features of the same object. To him, it was nothing like the way we learn to associate the appearance of a food item with its taste.

Recently Molyneux's question has been studied outside of philosophy, in science labs. In a 2011 study in *Nature Neuroscience*, two cognitive scientists at the Massachusetts Institute of Technology tested three people who were born without vision but then had their vision restored through a humanitarian organization in India called Project Prakash.[8] When the participants' bandages were removed forty-eight hours after surgery, they were easily able to touch a Lego and know what they were sensing. But they had a very hard time recognizing the same Lego piece when it was held up in front of them.

A few days later, after the participants had more practice relating the visual to the tactual, their ability to make these associations improved, which suggests that they had to learn to relate what they saw and what they touched to each other. It seemed Molyneux's question could finally be laid to rest: Locke had been right from the start. But then critics started to question the research methods. There was a mismatch between how the patients were introduced to the blocks through vision and through touch. In the case of touch, they were allowed to move the block around in their hand. But when they were looking, they saw only one angle. It's also unclear whether the subjects improved solely because their vision got better after more time without their bandages and not because of any improvement in interpreting their senses.

What we do know, regardless of the true answer, is that touch and vision operate closely together, and even if their cooperation isn't automatic, it can be learned pretty quickly. If we lose one, the other steps in. Blind people compensate for their sensory loss by using touch. Those rare people who lose their sense of touch are able to make do by using vision. But the debate over Molyneux's puzzle endures for a number of reasons. First, it's an accessible question that average people have sharp intuitions about. Second, there continues to be the hope that science might answer it eventually, unlike so many other philosophical questions. Third, it connects to our broader bewilderment about the senses and their relation to the truth.

The assumption at the root of it is that touch and vision are telling us about the same object. But there are times when they're not on the same page, as in virtual reality. When they're telling different stories, the brain is in an extremely uncomfortable position and tries to create a cohesive message for us by deciding which sense to believe. In the competition between our eyes and our bodies, our eyes usually win. Our tendency to believe what we see is largely because vision accounts for more of the brain's computing power than touch, with the visual cortex taking up about 30 percent compared with 8 percent for touch, although this doesn't include proprioception, which makes it an imperfect comparison.[9]

To convince us of the genuineness of these images, the brain silences messages it receives through touch. The brain has regular practice in ignoring these feelings, such as the friction from our clothes or the strain of our limbs, to help focus attention on what's new in our surroundings. This gives us a natural propensity to disregard it. What we see gets us caught up in another existence, and we don't realize that the more time we spend there, the less we notice our true embodiment. Although we view ourselves as firm and solid, our mental representation of ourselves is always in flux. It's scary for people to think that even our most basic truths could be illusions. Then again, a widely known fact about humans is our capacity for self-deception.

Our tendency to fall into error isn't limited to our sensory perception; it extends to our personal lives. Because the senses are often the subjects of metaphor, their relationship with each other is used in literature to describe how we can go wrong by trusting what we see or feel. One of the earliest examples is the old Sufi story about the blind men and the elephant. As it goes, a group of blind men each touch a part of an elephant and try to describe what it is. One is grabbing its tail, and one is grabbing its leg, and one is grabbing its trunk. Because they all come at the elephant from a unique perspective, their own subjective experience, they aren't able to accurately identify the whole. Touch is fallible, this story says. If these men had vision, they'd be able to step back and see the big picture.

In contrast, there are artists warning about the consequences of overreliance on vision and not enough attention paid to touch. For example, in the seventeenth century, the poet John Milton, who wrote controversial political commentaries, was ridiculed by his critics for going blind. They equated his blindness with ignorance. Milton fought back that his blindness could have been the reason he had a tighter grasp on reality than they did. It made him less likely to succumb to illusions and grounded him in a deeper truth. He wrote: "As far as blindness is concerned,. . . I would prefer, if necessary, my blindness to . . . yours. For yours, drowning the deepest senses, blinds your mind to what is sound and solid; mine, with which you reproach me, takes away only the colour and surface of things."[10]

The same theme appears in Shakespeare's *King Lear.* The powerful Earl of Gloucester is convinced by his conniving, illegitimate son to believe that his earnest, true son is plotting to kill him. The earl is a gullible man and begins a manhunt in revenge. Later in the plot, when the earl is blinded and thrown out of the castle to wander the countryside by some of his political rivals, he comes to recognize how mistaken he was. He understands that when he still had his eyesight, he was blind to the honesty of his son. "I stumbled when I saw," he declares remorsefully near the end of the play. It is only when punished with the loss of his vision that he can recognize this truth.[11]

Vision and touch represent the mind versus the body, the truths of our surroundings versus our internal realities, respectively, and so these writers are really speaking about how to consider these values in our lives. While our eyes do give us a broad, objective view, they argue that it can be a skewed one. When we're engaging mainly through vision, our locus of control is outside of us. We are increasingly susceptible to the stories our friends tell or to media and marketing, which are often just smokescreens, and we don't listen to our gut instincts, even though they might be screaming out in disagreement.

One of the most famous scientific experiments showing how this opposition between touch and vision happens within our bodies is the rubber hand illusion, discovered in 1998. A subject is seated with one hand in view and the other hidden behind a tall screen. In place of the out-of-sight hand is one made of rubber. When an experimenter strokes at the same rate both the rubber hand and the subject's hidden hand with the same object, such as a feather, the subject becomes emotionally attached to the one made of rubber. The rubber hand feels like a part of the body, even if the subject knows he is looking at an inanimate object. The trick to the illusion is that touch and vision are being activated in synchrony. The more congruous they are, the better it works.

The illusion is so convincing that if the experimenter threatens the rubber hand by, say, swinging a hammer at it, the subject will pull his hidden hand away, even though he knows logically that he's not going to be harmed. This is what I experienced when I found myself ducking away from danger in a virtual setting. The mind becomes so identified with the rubber hand—or, in my case, my avatar—that physiological changes occur. The hand that has been mentally replaced with the rubber hand will get cooler, a sign that it no longer has thermoregulatory control. Its histamine reactivity increases, suggesting that the immune system starts to think the real hand is the interlocutor. Literally, we lose touch with our true selves.[12]

Paul Jenkinson, a senior lecturer at the University of Hertfordshire, was part of a team that has recently done follow-up studies on the rubber hand illusion and found that particular types of touch are best at eliciting it.[13] Harried and anxious touch isn't as effective as a soft, caress-like touch. This made him and his team believe that the kind of affectionate touch that's likely to come from a loved one can help us feel more like ourselves. Caring touch could even be a way to keep us from getting swept up in illusions created by vision.

Jenkinson, with his collaborators, next tried out the illusion with anorexia nervosa patients. These patients tend to rate a caressing touch as less pleasurable than healthy peers, possibly due to a malfunction in their reward system that made them identify more with the visual of their bodies than their inner sensation. When they were given the rubber hand illusion, they were more susceptible to it than the average person. Their mental representation of themselves was more malleable because the feeling of their own bodies couldn't provide a counterbalance. Several small studies have since shown that using touch therapy on anorexia patients could make them less receptive to the rubber hand illusion by getting them to regain their awareness of their internal body states.[14]

One former anorexia patient, Rachel Richards, author of the memoir *Hungry for Life*, has reported exactly this phenomenon in her own recovery.[15] It happened when she joined massage school for a career change. By this time she was in her forties and had been struggling with eating disorders for decades and was already showing signs of early osteoporosis. Although the idea of getting undressed and having her classmates practice on her made her tense, she gritted her teeth and did it anyway. After her first massage in class, her body already felt more present to her. She started noticing small details such as the rate of her heartbeat or whether a conversation left her drained, feelings she felt she had for a long time been suppressing. More important, she learned to see her body as a source of pleasure again, and she began to trust what she felt instead of believing what she saw in the mirror.

"I'd spent a long time detached from my body, ignoring my hunger signals, my exhaustion," she says. "Once I started receiving loving touch, the whole system started working together again. My mind and body weren't separate."

This research could have important implications for our health. Anorexia isn't the only condition associated with a reflective disconnection from our feelings. Trauma victims may experience the anxiety of their bodies as a threat and learn to ignore them. Chronic pain patients' minds also trick them into rejecting their aggrieved body part. The portion of the brain devoted to representing the body part shrinks, which can prevent them from seeking attention they need. Drugs are another way we numb out uncomfortable feelings. Because of the skin's role as a mediator between what is outside and inside us, gently coaxing it helps to highlight all of the internal sensations we could be ignoring and get us to take action to fix them.

In some cases our natural partiality for vision can be exploited to change how we feel. According to Lorimer Moseley, a professor of clinical neuroscience at the University of South Australia, people with long-term pain in their hand have been found to experience more pain and swelling when looking straight at the hand versus when they aren't paying attention. If they look at the same hand through a magnifying glass, the pain increases.[16] When the hand is viewed through upside-down binoculars that show it as smaller and further away than it is, these patients have less pain. Although nothing is actually happening to the hand, these patients' brains can't help but create false narratives based on what they see.

One way this trick has been used in therapies is through the much publicized mirror box illusion, which is used to cure patients with phantom limb pain. Patients are asked to put their surviving arm through a hole in the side of a box with a mirror inside so that when they move the surviving arm, they are deceived by what they see in the mirror into thinking there are two arms that are functioning and moving normally. This, in turn, fools the brain into thinking that they're cured, and the pain lessens.

Gromala's virtual reality game is another example of this phenomenon at work. While she would like to make the therapy available to more people, as a pain patient herself, she doesn't want to provide any false promises. She is in the process of testing it to make sure it consistently works. She does this through what's called a diffuse noxious inhibitory controls test, in which participants dunk their hands in ice water and see the degree to which they can handle the cold. Pain is admittedly a tricky aspect of touch. It can tell us about objects external to the body, a pinprick or the chillness of water. But there is also pain that's purely internal, which most would say have nothing to do with touch. But they're both interrelated, which is why cold can be used as a proxy for pain. Gromala gives the test to patients before and after the simulation to see if it makes them any better at handling a freezing temperature. For many of them, it does seem to extend the time they last.

To show me how she conducts this research, Gromala brings out a bucket and asks me to put in my right hand for as long as I can. I'm not too worried about the cold at first. I still have memories of the Ice Bucket Challenge that went viral on Facebook and watching my friends douse themselves in ice water. I'm sure this will be nothing in comparison. But ten seconds in, I realize how cocky I've been. The water feels like it has teeth that are piercing my fingers with a hundred little holes. After a few seconds, my hand goes numb but my mind keeps sprinting.

Will this kill my nerves? Will my fingers fall off? I look at the bucket. It is red and green with a tugboat and a moose on it, not very scientific-looking. I worry that Gromala, as an artist, might not know about long-lasting side effects of this experiment. I recognize that I'm overreacting here. But this is the feedback loop of pain in action; my rumination increases the unpleasantness of the cold pain, which in turn triggers additional anxiety. I ask Gromala how long most people last in the water.

"Not that long," she says. "Less than thirty seconds."

I count time until I can claim that I'm above average, and I pull my hand out. I make it precisely thirty-eight seconds. I'm not part of a formal study, so we don't have a test from beforehand to compare to. But the results do make me wonder if I stayed any longer because I'd been living inside an avatar for a few minutes right before. If I'd worn my headset for another hour a day, I think about how many more seconds I could have lasted and how much more detached I would feel from all of the feelings of my body. As I think back on my virtual experience, I'm less focused on how quickly I merged with my avatar on the screen than on what feelings I might have suppressed for that illusion to occur.

So why is Gromala's work a stand-in for today's digital world? Within our highly visual culture, there are very real consequences for our sense of touch. Consider one of our most frequent activities: staring at our phones. It doesn't even matter that we might be in the midst of a crowded restaurant. We don't hear other people's conversations. We don't whiff the smells coming out of the kitchen. Everything around us fades out of focus. All of this isn't so surprising; it's often even our intention to remain aloof. Our phones are the armor we wear in public to let people know we're unavailable. But what we probably don't realize is we're cutting off parts of ourselves at the same time. Staring at that singular bright light, every other kind of sensory awareness ceases to exist. Even the most basic of our bodily functions don't work normally.

We don't swallow as often. We don't breathe as evenly. Our bodies go numb, and we become just brains floating around in space. We may not be entering an avatar, but we certainly aren't living fully inside ourselves. Although moment to moment we might not realize it, over time this disconnection leads to a tension we can't quite place. We know instinctively the value of touch and what we could be missing. We get nostalgic about the idea of working with our hands, and we dream about having closer physical relationships with our loved ones. We lament how out of step we are with the cycles of nature. And yet we ignore these feelings and continue scrolling and clicking. We come to see ourselves as projections and not as complete people.

In response to this cultural progression, an entire subgenre of articles has arisen on digital detoxing by writers who wonder if the belief they've been sold, that an active online life will make them more successful and happy, is actually leading to their misery. Their narratives are almost identical. The authors put away their devices and return to nature. Without the constant flashes of light calling for their attention, they become calmer. They start to regain other sensations, mainly touch, that they've been suppressing, through dancing or making crafts or simply walking, and that makes them more attentive to all of their inner needs that have gone unrecognized and unmet. These stories are about restoring a sensory balance.

In the article "A Trip to Camp to Break an Addiction," in *The New York Times*, Matt Haber attends the most well-known of all digital retreats, Camp Grounded, essentially an adult hippy summer camp, and writes about taking a meditative breathing workshop, "wherein I learned how to hug someone by positioning my head to my partner's right side so that our hearts could touch and our breath could sync."[17] He talks about napping in a hammock, dancing, and having his face painted, all of which represent a return to the embodied experience. In his essay "I Used to Be a Human Being" in *New York Magazine*, Andrew Sullivan writes about leaving social media to spend time at a silent retreat. "It was as if my brain were moving away from the abstract and the distant toward the tangible and the near."[18]

As readers, we know that time away from computers and smartphones doesn't represent a return to any kind of recent reality. It's not like we were perfectly in tune with ourselves before we had screens. We've long had other technologies that mediate the space between our internal and external worlds, and we had societal expectations before that. People have always existed on a continuum between acting out their innate nature and conforming to the needs of society. Social media is just the latest and most visible iteration of this conflict. But the underlying message of these essays rings true. The discomfort we feel when we hide parts of us, especially from ourselves, can be traced right back to the structure of our senses—the loss of touch in favor of vision.

This sensory trend has been of interest to Richard Kearney, a philosophy professor at Boston College, who refers to our time as the Age of Excarnation, as in living outside of our bodies. When I interview him over Skype, Kearney has the authoritative presence of a public intellectual. He has the pleasant-sounding voice of someone who could make a career recording audiobooks and is the kind of person who likes to break out into multisentence quotes by obscure philosophers. He lays out for me the seemingly infinite ways our technology makes us more distant from each other.

Therapists hold sessions online instead of on their couches. Surgeons perform operations digitally using robots. War is conducted remotely, with soldiers visually tracking their targets for weeks on camera before pushing a button to attack them. We communicate via text message more often than in person. Our social status is determined by likes on our social media posts. We ask apps to monitor our exercise, our stress, our sleep, our productivity, and our diets. What our bodies tell us about their own well-being matters less. Instead, we outsource these important functions to our technology.[19]

Online dating is a search for love operated mainly by vision. We decide our compatibility with a partner based on a set of images and descriptions. Regular online messaging and even negotiations about having sex often occur before a real-life meeting ever happens. There's a kind of empowerment that comes from implicating our images, rather than our bodies, in these decisions, which is useful for those who those who are reserved or whose sexual preferences are stigmatized and harder to reveal in person. We don't have to face the hurt of rejection. But there's also plenty of room for concealment or distortion of our true personalities, which makes the entire goal self-protection rather than a wish to be known.

When scrolling through an online catalog of people and projecting our wishes upon what we see, there's little chance to truly be vulnerable to each other. While we're free to touch, we want to remain untouchable. The benefit of a slow, humid crawl toward intimacy is that it allows us to feel each other out. We're thinking not just with our logical minds but with our bodies and instincts. There's a back-and-forth exchange that opens us up over time, which creates the feeling of mutuality. Additionally, the way people have sex is highly influenced by the performative example of pornography. Porn itself is a reflection of a culture obsessed with images. It is fueled by a desire to fulfill sexual urges without ever having to deal with the complexities of a real person. It's designed to be visually stimulating, not pleasurable in a true carnal sense.[20]

"What's missing is a ceremony around touching," Kearney says. "Caressing and courting and dancing, and that sounds very old-fashioned I know . . . I'm all for the Internet and what have you, but I think there is something lost, not that it can't be recovered. But when the body of the other becomes primarily mediated by vision, not touch, you're not exposed in the same way that you are behind a screen."

Medicine also has oculocentric tendencies. A doctor is asked to maintain a level of detachment when examining a patient and determining how to intervene with a cure. More often, she will use a scan instead of her sense of touch to make a diagnosis. It's been a long time since the days when physicians were also healers who tended to patients' pain through counseling and touch. The physical exam is one tactile ritual that has remained, although it too seems endangered. It's not just about checking lymph nodes and reflexes. It's a subtle signal of unhurried attention and care. It signifies a bond of trust between two people. The psychological impact is much more important than we realize. "It's possible to be a doctor who doesn't touch a patient, but it's not the best way," Kearney says.

Kearney sums up that when we cease to exist in our own minds as fully fleshed-out people, we're more concerned with appearance and external validation than with meeting our internal needs. In turn, we treat others with less sensitivity. If we recalibrate ourselves, he implies that many of our social values would change. We wouldn't try to be the loudest or most visible in the room but instead the most sensitive. We would prize subjective, emotional experiences over universal, observable truths. He thinks that in our move to a highly visual existence, we're denying deep truths about who we are and what we want, and we're probably making ourselves unhappy.

"It's about balance. Be with people in a way you can hug them or take their hands or touch their arms. It's not going back in time or tree hugging and that kind of thing. It's about recalibrating."

But he, just like the essayists, doesn't have practical steps that we can take to come back to a sensory equilibrium in the long term. While a call for small shifts—being present with people and hugging more—sounds easy, if all of the unconscious messages we've been getting our whole lives are telling us that touch is a base or unnecessary sense, then it's nothing short of radical. It involves unlearning what we know about touch, and we have to be constantly cognizant or else we'll keep slipping back to our usual habits. Not to mention, we don't know whether the people around us will even want to cooperate.

5

Softening

Overcoming Touch Aversion

With the context I now have, I'm ready to take a new approach to my tactile world. I start with at a place that's highly personal for me: my fear of touching other people. At the beginning of Western Massage 1, my teacher, Al Turner, a wiry man with glittering eyes who used to be a professional dancer, asks us to line up. He bends his knees, sinks his weight into his heels and sashays from side to side, a movement he calls "horse dance" and asks us to follow along. This is the kind of large, sweeping motion we'll use when we're giving a massage, he says. It gets us to engage our whole bodies, including the strong muscles of our legs and our core, so we make fluid strokes and protect the smaller, more fragile bones in our fingers when we're massaging.

Things get intense really fast. Next he has us pair up at massage tables to practice using this motion on another person. My partner, a Rubenesque yoga instructor named Elena, lies down first. Turner directs those of us who are still standing to touch various parts of our partners' bodies, first the backs of their arms and then their legs. He asks us to pay attention to how they feel, whether they are tight or slack, hot or cold. We have to monitor how they react, if they flinch or their breath slows. I'm not really noticing much of anything he wants me to. I'm mostly focused on my own awkwardness. As I go over her body with my fingertips, every snag in her clothing or bump in her skin is magnified in my mind.

"Remember to use your whole hand," Turner says. "If you're nervous, she's going to feel it."

He sounds as if he is speaking to everybody in the room, but the location where he's standing suggests that his comment is meant especially for me. I look around at all the other students. There are a couple who look sort of uncomfortable, but most of them are at work totally unfazed. I guess massage isn't something you study if touching people is hard for you, unless you're me. I feel Turner watching me as I place my palm down on my partner's back. My shoulders tighten, and immediately he pipes up again.

"If you're uneasy, you can calm yourself down by going back to horse dance," he says. "Remember, you're just dancing."

He tells us that when he's massaging his clients on days he's not teaching, he imagines himself cutting a rug around the table, just like he used to do on Broadway stages. He strongly believes that if we're having fun, the client will too. The positive vibes will seep straight into their skin. I close my eyes and imagine the Swedish popstar Robyn playing in my head. My heart speeds up, and I let myself become looser. The minutes pass faster, and before I know it, it's time for break. The first thing I do is run to the bathroom to wash my hands.

I'm telling you about my unease at massage school to highlight how, for a sense that we associate with comfort and pleasure, it also evokes my most extreme fears. I don't consider myself cold or withholding. I do like people, and I enjoy touching the one or two I'm closest to. It's one of the best feelings in the world. That's what makes me wonder why I don't seek it out more often and why, when people reach out to me, my first instinct is to recoil. I've tried for years to sort out my thoughts about it and assumed it's just a product of my shyness. But some research on touch avoidance shows that it can say much more than that about our personalities.

Because touching requires us to put ourselves out there, people with more confidence are likelier to initiate it when they're young. The positive reinforcement they get in response leads to a feedback loop that gives them an overall more open and expressive body language. People who are touch avoidant tend to be less comfortable in their own skin and could suffer from low self-esteem.[1] They are often passive and have high degrees of inner tension, meaning they have conflicting feelings, such as simultaneous desire and fear, that prevent them from taking action. Some have faced interpersonal trauma that makes their emotions feel particularly dangerous to them, which is why they feel two ways about acting upon them.

Some of these traits are indeed tied to what we think of as introversion, which is partly something we're born with.[2] As babies, natural-born introverts are highly reactive. They notice and respond to every sight, sound, and smell. As a result, they approach new situations with alertness and trepidation, which also means they need more downtime to be alone and process their feelings afterward. Babies that are less observant of every small change in their environment often grow up to be extroverts because as they grow up they aren't fazed by the little nuances of each exchange with another person. The way they use their bodies reflects their innate nature.

Of course, our temperament is only part of the puzzle. Upbringing also matters. If when we cried out as babies we felt our parents were highly responsive to our needs, we developed a secure attachment. We got the sense that our loved ones would be there when we needed them and assumed the same as we formed friendships and romantic interests in our later years. If we felt our parents were unavailable or aloof, we interpreted their lack of touch in various ways—that we needed to learn to be independent and soothe ourselves or that relationships we sought would always be a source of anxiety.[3] Depending on our experience, our preexisting traits were either tempered or heightened.

Those with insecure attachment styles generally take one of two routes: we grew up to be either somewhat avoidant or suspicious. People with avoidant attachment styles report having less enjoyment of emotional and physical intimacy, including touch. They may have taught themselves from a young age to suppress their need for affection, believing that being too demanding with a caregiver would lead to abandonment. While people who have suspicious or anxious styles of attachment still find touch rewarding, they could enjoy it less for other reasons. For example, they might be quietly conscious of how much their partner touches them and read too much into it. Time apart has the potential to make them feel unwarranted distrust. The positive side of closeness for anxious types coexists intimately with the fear of losing it.

The relationship we had with our parents keeps getting played out throughout our lives, according to what psychologists refer to as attachment theory. To be clear, our style of relating isn't a direct reflection of our parents. It's just our interpretation of our relationship with them. We may have assumed they were uninterested in us when they were just busy with their own problems, like managing a demanding work schedule or dealing with health problems. Or we may have taken our parents' behaviors personally when they were just acting out the patterns of their own upbringing or their cultural script. We could have interpreted as coldness the composed nature that their parents taught them to embody. Even when they're only partially true, the stories we tell are important because they provide a window into our behavior.

I was a natural-born introvert who basically couldn't have been more reactive. The first time my feet hit a sandbox, I threw a fit. My family once went on a trip to Disneyland that became a nightmare for everyone involved because I was afraid of everything I saw. I constantly had enormous feelings. I was also perceptive enough to notice how stressful my emotions were to my fairly nondemonstrative family. So I overreacted in the other direction. I decided at some point in my life that rather than reaching out for help, I should handle my emotions on my own. I didn't understand that I was stifling my natural impulse to reach out and cling to others for help.

Over the years, there were other lessons that added to my anxieties regarding touch. I heard many adults within my Indian immigrant community sharing aloud their fears that most Americans are insincere. It's a social script that's only natural for a group whose culture values quieter self-expression, suddenly surrounded by one whose main ethos is charisma. So I became suspicious when someone touched me. It's not that I didn't want closeness; I worried that it wasn't genuine or that the person giving it wasn't dependable, and I didn't want to get attached. As an adolescent, I learned from school and the media that innocent-seeming touch can be a sign of coercion and that men usually touch women when they want sex. As much as I tried to push myself to be more physically expressive, these beliefs were roadblocks I couldn't get past.

Everything changes as I get further into massage school. I'm unpeeling all those layers of programming. Class is about the same every week. We learn new parts of the protocol—how to drape the sheet and reveal only one body part at the time, how to massage the back, then legs, then front. The first thing I notice is how much more salient the sensations of my body become to me. I wonder how long my breathing has been so shallow and the muscles in my neck so tight. The twinge of pain I've had in my hips for years intensifies. I've clearly been ignoring for a long time all the tension I've been carrying. We're told it's a good thing to push beyond our limits, but in listening to my body I finally have a chance to heal it.

Once I can feel my feelings, I become better at telling my massage partner where my problem areas are and what exactly I'd like them to help me with. I've never before known my body or my preferences well enough to speak up. I'm also learning about consent for the first time in a classroom setting. It's a far cry from the sexual harassment webinars I have to take at work because now I'm actively practicing what I'm learning. As massage therapists, we're taught how much control we have when our clients are lying down prone, unclothed and how to be conscious of it. At each step, as we take turn giving and receiving massages, we have to ask our partners for permission to continue and remain communicative, checking in to see if we're doing okay.

About halfway into my semester, because I have to stand in line at a local copy shop to print out some massage homework due that day, I'm late to class. Turner is already fifteen minutes into a demonstration on how to massage the back of the legs and the stomach area. When it comes time for us to practice on our classmates, I look around for a partner. Everyone is already paired up. A guy named Ron, a personal trainer with muscles braided like challah and a handsome smile, asks me if I want to work with him. Until this point, I've avoided being partnered up with a man. I get nervous again.

I don't have any real opposition to it. It's just an instinct. As a woman, I've been told countless times to be careful about my interactions with men, and one that involves lying down unclothed seems particularly risky. I don't want to be prudish, so I say okay. Ron hovers a sheet over my body as I lie down on a massage table to let me undress. He looks away as I wriggle out of my pants, my massage school-issued t-shirt, and bra. I remember that I haven't shaved in a week. He undrapes my leg, wrapping the sheet around my upper thigh and securing it by tucking it under my sides, and I wince a little with embarrassment. He notices me tensing up, and he places his hands on my lower back, reassuring me. It's sweet of him.

That gesture makes me question what I'm so afraid of, especially in a roomful of people where I can speak up if anything inappropriate happens. Where did I learn to fear any man's touch? By this point Ron has moved on and proceeded to the exercise of the day. He fumbles and forgets at times, and I can tell he's more focused on getting it all down than he is on me. I calm myself by realizing he's a student learning a lesson, and I might as well be a textbook. I feel sort of silly for thinking of this exchange as anything else and relax under the sheet. Ron notices something has shifted in my body and stops to ask if the pressure is okay. The truth is he's being so gentle that I can hardly feel anything. I tell him I'm fine. I feel safe and heard and so comfortable that I let myself fall asleep.

It's a transformative day for me, as are the next few classes. Instead of thinking through how I'll come off or what other people are thinking when they touch me, I treat each massage as a matter-of-fact transaction, without any of the interpersonal dynamics that can make touch scary in the real world. I'm learning to give and receive touch for the simple enjoyment of it. I recognize that it doesn't make me soft or needy or dependent to enjoy being cared for by another person. I become more mindful of that part of myself deep within that secretly always wanted physical closeness and question all the assumptions and fears I've built up about it, including the belief that everyone around me would prefer that I remain at a distance.

I learned a long time ago that I needed to close myself off to stay protected. But by doing that I've lost one of the powerful ways I have to express myself. What I want to do instead is to trust that when someone shows their care for me, they might actually mean it, and to know that I have a right to remove myself if I feel otherwise. Outside of massage school, I find myself naturally reaching out more to touch a friend's shoulder or squeeze their hand. It's like I'm conversant in a new language that I've always known somewhere in the back of my mind but have been too nervous to mispronounce. I don't evolve into a different person. I'm not suddenly touching everyone I meet. But I open up to using touch to convey the warmth I always had inside.

It should be obvious that touch is important to our well-being. But, interestingly, this is something we've only recently come to accept about ourselves. Back in the 1930s and 1940s our culture had a decidedly icier bent. Because of new evidence for how bacteria and viruses transmit illness, doctors emphasized the value of proper hygiene, including preventing unnecessary interpersonal contact. In foundling homes, where abandoned children were sent, administrators accordingly lessened the amount they handled their charges and moved their beds away from each other, surrounded with mosquito nets, to prevent cholera. Mortality rates did decrease significantly, and the message became clear: sterility was the key to health, and touch was its nemesis.[4]

This lesson trickled down into home life since parents wanted to do everything they could to raise strong children too. The emerging field of child psychology was echoing what medicine said: touch makes children weak. John B. Watson, an American psychologist, said that excessive affection could produce soft, effeminate children who wouldn't survive in the tough conditions of their eventual workplaces. In his widely read 1928 manual, *Psychological Care of Infant and Child*, he urged parents to start training children early on for a world that would not conform to them. Among his bullet points: if children are crying, let them work it out on their own. Don't hug them. Don't kiss them. If you must, shake hands with them in the morning.[5]

"When you are tempted to pet your child remember that mother love is a dangerous instrument," Watson said. "Too much hugging and coddling could make infancy, unhappy, adolescence a nightmare—even warp the child so much that he might grow up unfit for marriage."[6]

Just as Pavlov had trained dogs to salivate by cuing them with the sound of a bell, Watson thought it was possible to condition children to having the kind of character that would be most valuable to society. It was just a matter of creating the right combination of stimulus–response associations. Watson's theories came up at a time when psychology was trying to gain a footing among the serious sciences, and his simple *x* leads to *y* algorithm was appealing for the design of experiments. His popularity coincided with the growth of a regimented factory system, which meant people were already prone to think of themselves as similar to the machines they worked with.

But there were a few holdouts in his field who argued that this was too simplistic a way of viewing human behavior. One of them was a young psychologist named Harry Harlow. He believed it was problematic to treat all human characteristics as a mechanical condensation of experiences and reactions. What these studies were leaving out were more magical and mysterious aspects of what it means to be alive, including wonder and love. He was convinced that part of the reason for this omission was that scientists were using far too simplistic animal models, such as mice and dogs as stand-ins for humans. As he visited his local zoo one day and watched the monkeys, each displaying a unique personality and engaging in multiple complex relationships, he realized that monkeys, as subjects for study, would be a huge improvement.

His lab became one of the first in the United States to attempt to raise rhesus macaques, which was the most affordable monkey available, so he proceeded with extreme care. He made sure they were fed well and took the proper medicines. To stave off infections, he separated them from their mothers and placed them in their own cages. Inadvertently, he re-created the same conditions of the foundling homes. As the monkeys got older, they started exhibiting odd behaviors. Racked with boredom, they sat and stared remotely into space. They sucked their thumbs and rocked their bodies violently. When they were brought together to breed as teenagers, they had no idea what they were supposed to do with each other. They weren't used to regular socialization, so they inched away to be on their own.

By all measures, these animals were well cared for. Harlow wondered what made them so mentally unstable. While observing the animals for hours at a time, the lab workers spotted a clue. Some of the monkeys had become absolutely obsessed with the cloth diapers that were used for lining on the floors of their crates. They hugged them and draped them around their bodies, almost as if they had an emotional dependency on them. This behavior was unique to monkeys raised in an isolated lab environment. Out in the wilderness, monkeys didn't seem to have the same kind of obsession with soft objects. The workers suspected that the warmth and padding supplied by the diapers could be replacing their need for something else important in their lives, such as their mothers.

This inspired Harlow to devise his now famous experiment. The baby monkeys he was raising were given one of two options of "mothers." One was a bundle of terry cloth, and the other was a wire frame. The cuddly mother proved far more popular than the rigid one. In follow-up studies, the researchers found that they had this preference even if the wire mother was equipped with a bottle carrying their food. They quickly ate and then ran right back to the cloth figure for emotional comfort. This flew in the face of the behaviorist belief that children become attached to their mothers because they associate them with nourishment. Harlow's lab had yet to learn how the simple act of clutching a parent could be developmentally valuable.

As the monkeys grew up, it was observed that those raised with the bundle of terry cloth approached new situations, like exposure to a teddy bear beating drums, with more courage than those with the wire version. When nervous, they lulled themselves against her soft surface and then proceeded to explore new situations. Those paired with the wire figure held themselves tightly or threw themselves on the floor and rocked violently.

They shielded themselves from what was unfamiliar to them. Motherly love, at least in monkeys, wasn't producing fragile children. Rather, a solid and stable connection to a mother figure helped them to form confidence that would foster their curiosity and independence as they reached adulthood.

By the 1960s Harlow was a household name, regularly giving magazine and television interviews about his experiments. He was a natural showman too. In fact, he was careful about reconstructing his terry cloth experiment with a cartoonish, smiling monkey head for the TV cameras so that a viewing public would see it as not just a bundle of fabric but a substitute for an actual mother. His promotion of motherly love and affection became the new ideal for parenting, replacing Watson's disciplinarian model. But, of course, the research he conducted was still just about monkeys, and it couldn't definitively be applied to other species.

A similar experiment on human children wouldn't have been ethical to conduct, and Harlow's work couldn't even be done under today's standards. However, in the mid-1960s, a true case of human touch deprivation sadly arose on its own. Romania's communist dictator, Nicolae Ceaușescu, instituted a series of policies to increase the country's population in hopes of boosting its industrial production. He placed sanctions on couples that didn't have a certain number of children and banned most abortions. While the plan did boost the country's output, it also backfired. The economy couldn't support these new, larger families, and many were forced to send their children to orphanages. There they continued to practice the hygiene-based standard of care from twenty years earlier.[7]

Videos shown in news broadcasts shortly after Ceaușescu's fall and execution depicted nonverbal, severely malnourished children who made jerky movements with their bodies. Most eerie of all were their blank, emotionless stares. That's how this issue caught the eye of Mary Carlson, a neurobiologist at Stanford. Carlson had studied with Harlow, and the children's behavior reminded her of his monkeys. In 1994 Carlson and her husband, Felton Earls, a psychiatrist, traveled to Romania to study these children. They compared the children in the touch-scarce orphanages with others from a small, one-year-long enriched program where abandoned children were given more caregivers who spent time holding and cooing at them.

Carlson and Earls found extreme differences in the stress levels in the two populations. Those in the standard orphanages showed irregular patterns of cortisol, a stress hormone connected with the fight-or-flight response, in their blood throughout the day, a pattern that correlates with lower performance on mental and motor development tests. Their growth was also stunted. The chemical functioning of children in the enriched program was much closer to normal, but unfortunately the results from this special program weren't long lasting. When they returned to their usual conditions, their old patterns mostly returned. As they got older, many of the orphanage-raised children were never able to leave institutions.

Around the same time as this research was going on, a then graduate student in developmental psychology at the University of Massachusetts Amherst named Tiffany Field started to look into how touch could be used to enhance the care of young children in medical settings.[8] She had recently completed a study that showed that giving premature infants pacifiers to suck helped speed up their growth. She reasoned that if stimulating the inside of the mouth was effective, touching the whole body would be even better. When her daughter was born prematurely, she massaged her body daily and found it calmed her down and got her to drink more formula. After she reported her anecdotal experience, the neonatal care unit where she worked did a larger study that replicated her results.

The then-CEO of Johnson & Johnson heard about Field's research, and he called her into his office and pledged $250,000 to help her found the Touch Research Institute in Miami. Since then the center's research has helped us to understand exactly how touch helps these children. When babies are born, their lives are filled with new stressors such as hunger, loneliness, and pain. Mostly nonverbal and immobile, they have to rely on their caregivers for nourishment and attention. Regular comforting touch assures them that someone is there for help and triggers a calming reaction. If they don't receive it, then their bodies react in ways that will let them survive for longer. They move less, and their metabolism slows. They stop growing. Their stress levels dial up so they can stay alert. These patterns, if chronic, can last through the rest of their lives.

There is a cascade of chemicals that make touch so important to our wellness. A welcome touch is shown to trigger the parasympathetic nervous system, which in turn sets off the release of serotonin, dopamine, and oxytocin—the triad of hormones that contribute to overall happiness. Prolonged deep touch, like that from a massage, can produce a large decrease in the hormone arginine vasopressin, which has been linked to aggressive behavior, and the stress hormone cortisol, part of the body's fight-or-flight response. Lowering cortisol in turn helps curb inflammation and increases the body's count of white blood cells, which play an important role in defending the body against disease.

The chemical brew that gets released is the opposite of what floods our bodies when we're living in fear and shame, emotions familiar to people who are lonely. Kory Floyd, a professor of communication at the University of Arizona, coined the term "skin hunger" to describe this condition. In 2014 he published a study of 509 participants from the United States and sixteen other countries that found people who reported having a dearth of physical affection also said they experienced more loneliness, depression, and mood and anxiety disorders than the other subjects of the study. They had more immune disorders and had worse health generally. The touch-deficient subjects were also prone to having alexithymia, which impairs their ability to recognize what their emotions are and makes it harder for them to develop close, secure relationships with others.[9]

All of this research on the health benefits of touch has led to some major changes in our culture. Sick babies were once sequestered away so they wouldn't be exposed to infection, but they are now allowed visitation and touch by their families. Dozens of neonatal intensive care units around the country use touch therapy to help promote babies' growth. In the past decade hospitals in the United States have increasingly encouraged skin-to-skin contact immediately after birth, finding that it regulates babies' breathing and heart rate and stimulates hormones necessary for women to breastfeed. These same hospitals tend to house babies with their mothers during their stay, touting the benefits of close proximity starting early in life.

Touch isn't just limited to newborns. Deep touch therapy has become one widely accepted treatment for autism in children. Recent research has found that many autistic children have impairments in sensory nerves that cause them to overreact to tactile stimulation or to fail to recognize it as rewarding social behavior. Specialists can train these children in strategies for reintroducing touch into their lives, either by letting them know how they can request a hug from caregivers or find strategies to comfort themselves. Gentle stroking can agitate this population of patients, but a tight squeeze, like a hug given predictably by a trusted provider, can be reassuring.[10] Temple Grandin, the respected animal behavior expert who has autism, famously built herself a hug machine, a box with pads on the side that she could make give her a tight squeeze if she pulled on a string.

Most parents today understand that children need affection in order to thrive. Nevertheless, many might be unable to provide it. It's not easy to consciously force ourselves to change attachment patterns that were rooted in us in childhood, as hard as we try. We are likely to pass on our own attachment styles. As our culture pushes for us to improve, at times it has swung too far in the other direction, guilting mothers who aren't able to spend as much time bonding in the early stages because of recovery from C-sections or are busy with work. Comforting touch is good for children, but no parent can do it all. We're a complex and adaptable species, and there are multiple factors contributing to a child's health; if they're lucky, they have more than one place to seek the kind of nurturing they need.

Where we could improve is in creating more of these opportunities. When we don't have the chance to receive touch in our lives, we neglect to learn what should be obvious—we should aim to feel good in our bodies. Starting at an early age, schools could give children simple exercises that help them connect with their need for physical contact. Teachers can have them lie down and pay attention to the rising and falling of their chests. They can stretch their arms and legs or rub tennis balls over their limbs and feel the changes across the skin. By being asked to pay attention and talk about these sensations, they can be reminded that their bodies are a source of pleasure. Concurrently, they should be taught consent. Across cultures many parents force children to express affection and accept embraces they don't want. Instead, we should be giving them tools for expressing how they prefer to greet someone—with a smile, a high-five, or a hug.

By the time we're teenagers, most of us have heard repeatedly that our complicated feelings and urges make us act irrationally. We're actively told not to trust our bodies, and what we need to be reminded of instead is that our bodies carry an important intuition that we should listen to. While teenagers have plenty of opportunities to play sports, which are mostly a way for them to look better and achieve goals, there's a dearth of ways to develop somatic awareness. Yoga or dance or manual therapies can remind teens to come back to the feelings inside them. This comfort we have in our own skin helps us when it comes time to negotiate the terms of our first sexual encounter or discuss our preferences with a partner. Without it, we're more willing to let others make these decisions for us.[11]

When we are adults, our behaviors are so deeply ingrained that it's hard to reverse them. Depending on our life histories, we attach different meanings to touch, which means our primary way of thinking about touch is as a form of communication—one that can be useful for good but also for harm, which it is of course. But beneath all of its meanings, it is a biological need. Closeness to others and the movement of our skin are as crucial to our health as exercise and a proper diet. It takes some effort for us to see it that way again. Particularly those with trauma could use some form of rehabilitation to become more comfortable in their skin and to work through the reasons why touch, something that we desire so innately, has become intertwined with pain.

This happens through finding safe spaces to give and receive touch, like I found at massage school. For some, that could be with a partner. For others, it can be through taking a friend's hand or body-centered treatment from a professional.[12] The point is to engage in touch for the sake of it, without any other agendas, which can reveal just how many of the stories we tell about it are false and limiting. Some of us may have absorbed that we don't need touch or that it's less important than sex or that it's used to assert power or that we shouldn't touch someone of the same gender. The reason it's so difficult to change how we touch is that we're mostly unconscious of our tendencies, so we continue to play them out without knowing it. When we receive a touch, we pick apart its intentions. But if we are always reading between the lines, we don't enjoy it as much. Without these thoughts, we can again feel at peace with our desire to be close to other people.[13]

Turner probably remembers my hesitation at the beginning of the semester, or he wants me to prove I'm as dedicated a massage student as those actually planning to be certified. For my final exam in Western I, he initially pairs me with one of my regular massage partners, but then he changes his mind on the spot. His new choice is a Haitian woman in her forties named Lucrezia with kind eyes and who likes to refer to everyone by their nationalities. I'm "Miss India." From across the room, I've noticed that she has severe psoriasis. The large plaques cover her entire body.

When we meet at a massage table, she tells me not to worry, that she's not contagious, and I feel bad. I try to imagine what it's like to be in a body that makes other people so afraid she has to provide a disclaimer. Some people with her condition might wilt and withdraw. I'm impressed how she puts herself out there, no matter what the judgment might be. As I knead her muscles, I try to communicate through my hands what I'm truly feeling inside, that I'm at ease while working on her. Turner comes around the room observing our technique, making comments on a notepad. But I continue on, ignoring him. I've progressed since that first day. Steadily, I make it through the entire protocol, and when it's Lucrezia's turn to give me a massage, it's one of the best I've ever received.

Once our final is over, we talk for the first time all semester, and it's one of those conversations in which we both tell each other the entirety of our life stories that led us here. She tells me how she learned traditional medicine from her grandmother and had used it to cure loved ones who weren't helped by Western medicine. She believes she has a God-given calling to be a healer. As an agnostic, this is the kind of encounter that usually makes me sheepish. Normally I'm the kind of person who would be reaching for my phone if someone came and spoke to me about their otherworldly gifts. But in massage school, things are different. We joke together about my lifetime of nervousness about showing affection.

When feeling people's bodies and noticing where they're holding the tension, it's impossible to avoid getting into conversations about some of the more serious circumstances of their lives, even the esoteric. Because of that, my classmates and I have quickly formed a tight little community even though we seemingly have little in common. There is a retired ballerina, several fitness trainers, an elder care worker, a photographer, someone thinking about quitting his office job at a university, and a guy who trains wild animals. They come from different countries and socioeconomic backgrounds, and there's probably no other way besides massage school that we'd all be in the same room. And yet I don't know if, in my adult life, I've ever gotten as close to anyone so quickly.

Most of us meet up in the break room to talk about the exam. Some of us admit to forgetting parts of the routine. Others muse about what Turner could have been writing down. We all think we did okay. At least, we assume he would have pulled us aside if we'd failed. I tell everyone about a memory that sticks out from the semester. It happened one week when I'd come to class with a pain in my rib from a snowboarding accident. I came to class but told our teacher I could only observe, not participate. During break time, several classmates hovered around me, asking if I was doing okay; they wanted me to show them where I was hurt. One of them had me lie down, and during the few free minutes he had free that day, he actually made me feel a lot better.

"That's not something that happens outside of massage school," I say. "That week, when I went to work or met up with friends, I might have gotten a sad look but then everyone sort of moved on. I couldn't believe how much everyone wanted to help me."

A classmate named Rich practically bursts out with his own sentiment about the class. He's not actually in the massage program either. He's learning acupuncture and took Western 1 for an elective credit. But he's been so floored by the way that massage can change the dynamics between people that he's thinking about continuing. He feels more connected to the students in our class than he has ever felt in a school setting. Being aware that one of us might at some point be working on his body, he has had to drop all of his barriers. He believes the class has even made him kinder.

"I go to restaurants now and I look around at the other tables to make assessments about whom I'd get along with," he says. "Then I think about if I massaged them, how that would change."

Melissa, a French woman who wants to specialize in prenatal massage, says that outside of class she has become more affectionate with her family. She's always been that way, but she now feels that something within her has been unleashed—she can't keep her hands off of them. We've all gone through inner journeys along the way, even those who came in more comfortable than I did. Massage is about much more than relieving tension and getting rid of knots. It transforms our relationship with our own bodies' desires, which can help us open up to others' too.

Dozens of studies have shown that we just seem to like people who touch us. There's a higher chance we'd give a big tip a waitress who has touched us lightly on the shoulder, hand, or arm. Teachers who ask students to volunteer to solve a math problem on the board get more takers when they briefly touch them on the forearm. In Stanley Milgram's famous experiments, in which participants were asked to apply painful electric shocks to a test subject who gave the wrong answers on a series of word-pairing tests, compliance with the request reduced dramatically when the person who was given the directive was asked to hold the recipient's arm. Even in our touch-phobic society, many of us feel instantly closer to someone when we touch them.

While working with a team at Berkeley to test how touch contributes to group unity, Michael Kraus, an assistant professor of psychology at the University of Illinois at Urbana-Champaign, decided to look at sports. He predicted that more early-season high-fives, butt slaps, and fist bumps could predict cooperative behavior and better performance as the season progressed. He watched videotapes of early games of all thirty NBA teams during the 2008–9 basketball season and scored the nature and duration of touch between players. He also counted and rated instances of cooperative behavior between players for these games through behaviors such as passing the ball, setting a screen, and cheering on teammates.[14]

When he looked at these tabulations alongside performance statistics by the NBA, he found that teams that touched more showed better performance at both the individual and team level. However he worried that there might be limitations to this simple correlation. It was possible that players who were more skilled, and therefore more optimistic about their performance during a season, were also more prone to physical displays. Or teams with better ratings in the polls could play up their demonstrativeness for the media. But even when he applied statistical corrections for both potential errors, he still found that early-season touch translated into better performance later.

There's some evidence that the benefits of touch extend even more widely, not just to teams but to entire cultures. According to research done by Field's Touch Research Institute, cultures that use plenty of physical contact are less violent. For example, French teens are more likely while sitting together at a McDonald's to lean on each other, rub each other's backs and give hugs. American teens, in comparison, engage in more self-touching, such as playing with their hair and cracking their knuckles, as well as pushing and hitting. She believes that Americans' relative dearth of affectionate touching could be the cause of our higher rates of bullying, abuse in relationships, and general societal violence, including the use of guns.[15]

In a 1976 paper that the Touch Research Institute often cites, the American developmental psychologist James Prescott found that within nonindustrial cultures, those that were more egalitarian and friendly had parents who tended to lavish physical affection on infants. Low-touch cultures tended to have more theft and violence on a wide scale. Prescott's theory was that a lack of stimulation in childhood could cause them to seek other, more extreme forms of it in adulthood. We can't reduce the cause of violence down to one specific behavior, but taken together, the research presents a compelling case for the ability of touch to promote harmony.[16]

All of us massage students live in New York City, which is home to a lot of touch deprivation. It's a place where many people live alone and work around the clock and don't have much time for a romantic relationship, or do but just don't have much time to be together. Even though we're surrounded by others on the streets and in subway cars, we still avoid each other as much as physically possible. There was something electrifying about going to massage school and connecting with strangers. By allowing myself to touch people more—and exploring how it made me feel—I became more aware of my physical needs. Massage school was the exposure therapy I needed to overcome the emotional and physical discomfort that had plagued me for so long. I desired the affection of other people more than I once knew.

6

Boundaries

Knowing Good Touch from Bad

Sitting at my kitchen table with a group of professional cuddlers, Indigo Dawn tells us about a favorite client. The young man, who was a recent college graduate, was terrified of getting close to a woman. When he was interested in someone, he was too shy to move his body closer to hers or place a hand on her shoulder. If he did make a move, he worried he'd give the wrong impression, like that he wanted to have sex. As a devout Christian, he knew he wanted to wait until marriage. Privately, he would watch movie and video-game scenes in which couples hugged after a long time apart, and he fantasized about how magical that must be. He dreamed about finding someone to experience that with.

He sought out Dawn because it felt safer to give and receive touch in a professional setting where there were no emotional expectations and he knew that sex was off the table. Sometimes during sessions he would bring in a movie soundtrack and ask Dawn to reenact some of his favorite scenes. Dawn, who is in their twenties and whose assigned gender at birth was female but now identifies as genderqueer and uses the pronouns they/their, didn't recognize the scenes but remembers having to say lines such as "I'm so happy so see you," before giving him a hug. The exercise gave him a chance to finally experience what he'd been romanticizing for so long. He saw that when the boundaries were clearly laid out and respected, he was more comfortable. And after several rounds of practice, touch was less intimidating.

Several months into their professional relationship, the client thought he was ready to use what he'd learned out in the dating world. He mentioned to Dawn that he'd met a woman he liked, and they talked together about how he could approach her. At the next session he was happy to report that the woman liked him as well. Over the next few weeks they agreed to be in a relationship, and he realized that he no longer needed the services of a professional cuddler, although he credited that experience for helping him to get to that point. With Dawn he had learned that no woman was ever going to fulfill all of the expectations of the cinematic moments he'd built up in his mind for so long. Instead, he just needed to try to connect and be present and wait for a relationship to unfold naturally.

"I'm super happy that I'm able to help people in that way," says Dawn, who is pixie-like with clear gray-blue eyes.

Dawn has been a practicing cuddler for a few years, since completing a training course through a website called Cuddlist.com after learning about the profession through a friend who was trying it out. By that point, having been involved in an experimental community called New Culture, which aims to change the way people relate to each other through radical honesty, an emphasis on sustainability, and exchanging consensual nonsexual touch, Dawn wasn't fazed by being so close to strangers. It seemed inconceivable to be able to make a living doing something that was so inherently enjoyable. Although Dawn didn't realize it at the time, cuddlers are a growing field, and as public acceptance of cuddlers has grown in the past decade, some psychologists have even begun recommending them to their patients.

Dawn sees cuddlers as educators and counselors, not just a salve for people's loneliness. They look at people's scripts about touch and how they can be modified. We are all affected by social norms around touch. Dawn's own family, while loving, had the attitude that "touch equals sex," and so from the age of six or seven onward, Dawn received very little physical affection. It wasn't until Dawn got older that they even realized it was a need they had. Another cuddler sitting with us, Don Shanks, who is in his sixties and works as a government contractor, grew up in a fairly touchy-feely family. It wasn't unusual for his family to hold hands while sitting around watching TV or to press up right against each other while eating meals. But somewhere along the way he still picked up the idea that contact with his wife was inherently sexual.

It was only when he became a professional cuddler on the side that he realized this; as he has made an effort to change his behavior and add more platonic touch into their life together, their relationship improved. Shanks says that many of his clients are middle-aged women, and they are used to playing a caretaking role in their families. They come to him because they're exhausted and don't feel seen, even by themselves. This is why it can be a powerful exercise for them to practice answering questions like What would feel good to you today? and How would you like to get started? Sometimes it can take them several sessions before they even make a suggestion. Because women are often taught to play a submissive role and adjust to others' needs, the purpose of this exercise is to get clients to ask for exactly what they want, not just with a cuddler but also in their daily lives.

"I didn't have much empathy compared to now because I wasn't thinking about it," says Shanks. "I inhabited a very male world. This work is about absorbing someone else's life journey and reflecting that and accepting it."

Dawn and Shanks tell me that touch is a limited resource in American society, and depending on our identities—our sex, gender, and age as well as race and social status—we receive more or less of it. The sex and gender differences are particularly pronounced. It's more acceptable for women than for men to touch each other because women are praised for being nurturing, so they typically grow up receiving more of it than men do. Men, on the other hand, are socialized to avoid touching each other except in the context of sports. That means they're unlikely to receive any touch if they're not in a romantic relationship.

In heterosexual pairings, men are expected to be the initiators of touch, which is a signal of the power they're expected to hold. This creates an uncomfortable dynamic between men who are expected to dominate as a way to display affection and women who have to navigate these mixed intentions. To complicate matters, men don't have a lot of practice before attempting physical contact with women, and so they don't always know how to go about it. Women are taught from an early age that intimacy isn't safe for them, especially with men, and many women who come to cuddlers have faced trauma in the past. Conversations about the way we carry our bodies and the role of privilege have become more prominent in our media narratives over the past couple years. We're finally having public conversations about boundaries and consent, which is the beginning of a course correction in how we touch.

While these changes are important, they often lead to confusion and exasperation on both sides. As a result, touching other people has become even more complicated than it's always been.

Even though we've reversed our thoughts about touch in the nuclear family and recognize how important it is to childrearing, in public settings there's a growing ambivalence about how we should engage with each other. There are so many unspoken expectations we have to navigate that sometimes it seems easier to distance ourselves. In addition to our personal stories about touch, we're also carrying around this cultural baggage. That's perhaps why these days more and more of us are willing to pay for it.

To understand where we are now, we have to go back to when such conversations about touch began to emerge, which is about the time of Harry Harlow's groundbreaking research. There were mixed responses by women to what he said about the importance of maternal touch to development. For some, Harlow's work was an opportunity to reframe female bodies as powerful and life-giving, a contrast to their portrayal in much of religious literature as weak or dirty or dangerous. For others, it added to a growing list of ways that women were expected to be perfect mothers; touch became another thing they had to do exactly right. Many second-wave feminists found his message threatening. They worried that all his talk about maternal care, including the value of breastfeeding, could be used to shame them into accepting their traditional role within the family at a time when they were finally making strides in the workplace. "In contrast to feminist work, most of this popular and scholarly writing relegated women to the role of mother," writes College of Staten Island sociologist Jean O'Malley Halley in her book *Boundaries of Touch*, explaining why women were reluctant to accept work like Harlow's.[1]

Rather than focusing on how touch could be used to nurture, many feminist leaders were more interested in highlighting the inequalities in how men and women touch each other, particularly when it comes to sex. For decades the overarching attitude was that love is a spiritual affair for women and that they only engage in the physical side in exchange for companionship and cuddling. Increasingly, women believed that, just like men, they should be allowed to have casual sex purely for pleasure without being punished or judged for it or thought of as unfeminine. This sexual revolution occurred alongside the introduction of the birth control pill and the new scientific understanding of the female orgasm.

Many women felt empowered to consider for the first time how their lives could be more sexually satisfying. But, as expected, there were many remaining questions about how imbalances that women faced throughout their lives presented themselves in the bedroom. Some worried that the emphasis on sex was just allowing men to get more from women without having to commit to them, which was made worse by low wages that made it difficult for women to live independently, and that it pressured women into sex lives they didn't want. The activities that women have stereotypically been said to desire—emotional closeness, intimacy and cuddling—also took a backseat.[2] The result was that, for both genders, the idea of being soft and nurturing seemed like a sign of weakness, and being sexual was a sign of liberation.

There was also increased discussion of how touch could be used to harm. Activists brought to the public's attention for the first time ills such as rape, domestic violence, and sexual harassment. These issues had immediate resonance with the public, and movies and novels began to deal with the long-lasting effects of negative touch in people's lives. Antirape and antibattering movements took off. Women formed support groups in which they could tell their private stories for the first time. The public came to see just how frighteningly common these experiences are and how they needed to be tackled as a national problem; after all, the way that men wield their power within their families reflects how they use it within public institutions when designing laws and policies.[3]

This led to an increased focus on another problem, which is abuse outside of the home. The phrase "sexual harassment" was introduced to the public's vocabulary in 1975 by a group of women at Cornell University after a former employee claimed she was inappropriately touched by a supervisor.[4] The school hadn't let her transfer positions and then refused her benefits because she had quit for "personal reasons." Even more women were speaking up about the behavior they were forced to endure in the workplace in order to keep their jobs, including sexual coercion. Within the next two decades, as prominent cases, including Anita Hill's, were publicized, American institutions did what they could to address this problem.

Companies created bylaws on what constitutes appropriate and inappropriate contact at work. Schools and districts developed codes that prescribe when it's okay for teachers to touch students. Sexual harassment training became ubiquitous at workplaces. While ostensibly the training was meant to fix a societal ill, it was more concerned with reducing a company's liability. A component of a number of these programs are lessons on avoiding touch, which push the notion that an innocent squeeze of the shoulder or pat on the hand could be misread by the recipient or, worse, escalate inevitably into exploitation.[5] But demonizing touch has not stopped abuse, which is because, first, a short session is unlikely to prevent a harasser from acting out and because, second, evidence shows such trainings can amplify preexisting biases about gender.

Even as there are more reports of sexual harassment, the consequences haven't risen at the same rate. Rules don't mean anything if they're not enforced, and the #metoo movement has revealed just how often the system fails to hold offenders accountable. Because institutions, including the courts, haven't always been effective at handling these complaints, we've mostly resorted to working them out ourselves.[6] But we have vastly different reactions to reports on inappropriate touch, depending on who we are and what our background is. Some of us might think that all of these strict rules have gotten burdensome and are only adding to our touch deprivation. Others of us find the rules important and necessary, especially if we've faced abuse. And people who grew up without much touch often don't understand other people's need for it, and vice versa.

We're caught up in a swirl of different philosophies and preferences. As I ask around about people's experiences navigating this confusion, several women talk about times a male boss has put a hand on their shoulder or the small of their back, which in most cases they read as a power move rather than a sexual advance that he wouldn't have used with another man. A few mention colleagues who have tried to make a move on them. None of them reported their cases to human resources. Even though they do have memories of a squeeze of the hand or a pat on the back giving them encouragement at brief moments, they don't see those as such a great loss if it's a way they be treated the same as their male peers. Most advocate a strictly hands-off policy at work.

Most of the men say they are fearful about touching colleagues, which means they flat out avoid it. One mentions a time when a photographer had to force everyone in his quite reluctant office closer together so they would fit into the frame. A male high school teacher I spoke to was accused of patting a female student on the butt during soccer practice; he swears it was her lower back. Although her mother took no formal actions, it was an important lesson for him. On the sports field, he'd always assumed that that it was okay to act a bit looser than he would in the classroom. He used to give bear hugs and wrap his arm around a student who scored a goal, but he is now careful to have a more professional demeanor.

But not everyone wants to follow these new norms. A former colleague of mine was appalled after he was instructed in a human resources training session to never hug his students. A boisterous, friendly man in his fifties, his approachableness has always been part of his appeal as a professor. If a student needed his affection, he wasn't going to deny it to them because of some policy. I watched as several of his female colleagues nodded in agreement as he expressed his disdain for human resources trainings on touch. They felt it was their role to encourage their students, and touch was an important tool. They believed that touching actually helps to break some of the power imbalances that make it difficult to learn.

It's not just at work that these kinds of negotiations happen. I talk to a mother who sends her children to a daycare that praises only students who are quiet and keep to themselves. Her daughter is outgoing and likes to go around saying hello and giving hugs, and she's worried that she could get the impression that her personality isn't valuable too. In a society that values extroversion, she worries that her child is learning that introversion is a sign of being well behaved. She might learn to suppress her naturally loving nature.

Dating is perhaps where touch is the most fraught. While the desire for touch and sex are intertwined, many men and women say that touch is the harder one to initiate. Casual sex is well-accepted, but a more serious relationship is required for cuddling, leaving many confused at how our norms have become so backward. Many women, but a few men also, say that they would be much more comfortable having sex if there were more physicality in the relationship leading up to it. Still, they don't break from expectation because they're worried the other party will think that affectionate touch implies wanting commitment immediately.

There are occasions when these politics of body language become part of the national discourse. That was the case when Michelle Obama put her arm around Queen Elizabeth even though it was a breach of royal etiquette. In her memoir *Becoming*, Obama tells the story about how they were both at Buckingham Palace, looking at each other with pained expressions as they complained about their uncomfortable shoes. The queen broke out in laughter, and Obama placed her arm around her. She wasn't aware at the time there was any rule against touching the queen.[7] The moment cemented her title as "hugger-in-chief" and was mostly analyzed as a triumph of American friendliness over stodgy British social norms.

While Obama's touch was a sign of breaking down barriers, Joe Biden received decidedly less pleasant reviews in the weeks before he announced his candidacy in the 2020 presidential election. A woman named Caitlyn Caruso said that when she was a college student attending an event on sexual assault, he rested his hand on her thigh even though she was squirming to hint at her discomfort. Later he hugged her for what seemed a little too long. Another woman said that he moved his hand down her back in a way that made her uneasy. Biden attributed his behavior to changing times, meaning that what had been acceptable in the past is being reconsidered in today's world. He described himself as a "tactile politician" and said he was listening and could adapt his behavior.

Following the allegations, we all projected our feelings onto Biden, whether we've been the recipients of unwanted touch or have been rebuffed. Men who struggle with confidence and who already have trouble making a move were worried this public backlash would make it even harder. Women who had accepted a touch they had not wanted finally felt that they could speak about it freely. Almost every man has been in the position of being nervous about touching a woman, for fear of being misinterpreted. Almost every woman has inwardly recoiled at an unwanted exchange. We heightened the narratives that expressed our personal desires for or discomfort around touch.

It was a moment for airing all the assumptions and dichotomies, real and inaccurate, that we have about some of our most basic human gestures, and the arguments grew even more intense when the allegations became more serious. As we try to correct for the patriarchal behaviors of the past, we expect those in positions of power to act with more sensitivity. Hopefully, men can understand that women's life experiences can make touch invasive for them, so men should make efforts to read women's behavior or ask explicitly whether touch is welcome. Asking "Can I hug you?" feels a lot more natural after saying it a few times. Women could also examine what makes them fearful of men's touch. It's possible that our past experiences have made us too quick to label any touch as bad. We could also make efforts to express in the moment what we want, now that we're being given more space to do so.

We're all better off when our boundaries are clearly stated. The only way we can improve is by risking and fumbling through engaging with each other. But even if we do take important steps—such as communicating what we want and becoming used to hearing the word "no"—it's important to acknowledge that touching in public life will always carry some risk. Human touch is as complicated as every other kind of communication we have, and there's no way to get it perfect all the time. It makes sense that the word "touch" in French has as its root the old French word "*touchier*," which meant to hit. As a noun, one of its definitions is a "slight attack," like with an illness; for good or bad, it can knock us off our feet.

In his essay "Skin Blind," in *Prison Masculinities*, a collection about how upholding a sense of maleness affects the lives of the incarcerated, a writer named Dan Pens describes the isolation he felt while imprisoned in the state of Washington, where there were strict no-touching rules.[8] These rules didn't have to be written down, he says. They were already self-enforced, even though everyone was living in close proximity. Brushing up against someone's shoulder by accident without apologizing could lead to lasting resentment and a future fight. Pens had been there for six years when he participated in an "Alternatives to Violence" workshop facilitated by volunteers from the community. One of them approached him afterward to thank him for participating, and he expected her to shake his hand. Instead she hugged him.

"It felt like warm, honeylike energy was being poured into the dusty, dry empty tank of my soul," he writes. "Touch is life. It is vitality. It is the music of the skin. It is the color that a blind man can't see. To be deprived of touch can wreak devastation on the psyche. . . . Our skin sings the song of life."[9]

He had never thought until then about how long he'd been deprived of human contact. It was something that crept up on him slowly, long before prison restricted him to the occasional bump in the hallway or on the basketball court. His deprivation might have begun while he was on the streets, when he was so numbed out from drug and alcohol use that he wouldn't have noticed it. Perhaps it was only because he had become sober in prison that he was able to feel the weight of human touch so deeply. After the workshop, he asked his cellmate whether he missed touch. He said he did because he was a very touchy-feely person. Pens considered giving him a back rub or a hug every now and again, but instead he kept to himself. Prison is a place where people keep themselves locked up, he concludes.

The setting of a prison is an extreme example of how men are socialized to touch. But while social changes have made women freer to ask for more space, men are bound by a narrative that requires them to be distant. Toughness and strength have long been valued as male traits, which has generally made men less affectionate. But even still, male affection was once much more common than it is today. Historian Joanna Bourke has written about how, in times of great change such as during the First World War, men cared for each other in sickness, bathed together, held each other close when dancing, and kept each other warm under the same blanket in the cold.[10] There are many historical portraits of American men up until the 1920s posing with their friends while holding hands, hugging, or sitting on each other's laps.

It was only as the twentieth century progressed and homosexuality became recognized as an identity that politicians and ministers could warn against and psychiatrists could label as an illness that men became cautious about avoiding the appearance of being gay.[11] Male friends avoided being close to each other, sitting close, or holding hands. Along with adopting a rigid body language, they suppressed or hid their emotional intimacy as well. The one exception has been in the context of sports, the subtext being that physicality is only okay in an aggressive context. A butt slap on the football field is considered acceptable;

that exact act in a locker room would be an inappropriate sexual advance. In other cultures, even ones where contact between men and women is tightly prescribed, it's not unusual for men to show intimacy.

Because touch between men is considered taboo, the one place where men can receive affectionate touch is in dating and other romantic relationships, but even that is becoming a scarcer prospect because we're not coupling up at the same rate. Fewer people are getting married, and those who do are choosing to do it later. About 60 percent of adults under the age of thirty-five now live without a spouse or a partner, which means we're collectively spending more time alone than before. There are numerous reasons we can cite for the loosening of our family structure, from economic uncertainty to changing expectations to overreliance on parents. But even among those who are in relationships, we're probably touching less. Sex is perhaps the best proxy we have since affectionate touch isn't studied nearly as much, and studies show we're having less of it.

In an analysis of the 2016 General Social Survey, conducted at the University of Chicago, Jean M. Twenge, a psychology professor at San Diego State University, found that millennials are having fewer sexual encounters, on average, than the two generations before them. They're also on track to having fewer sex partners. About 15 percent of people in their early twenties said they haven't had sex in adulthood. Across the board, adults have gone from having sex about sixty-two times a year to fifty-four.[12] These numbers have made the media panic about the possibility of a sex recession. Culprits include dating apps, the Internet, sex toys, pornography, and psychiatric medications, along with economic and social factors that have led to fewer marriages.[13]

It is worth mentioning that this is not all bad news. While previous generations may have viewed marriage as mandatory, today it is an option so we don't feel as compelled to enter bad companionships. Those who do opt for commitment wait longer, and it appears we're making smarter decisions. Marriages formed from the 1990s onward are more likely to reach fifteen years without a divorce. This is also a period when sex has been declining. What could be happening is that as women gain more agency, they're opting to have less sex. According to Stephanie Coontz, director of research and public education for a nonprofit group called Council on Contemporary Families, women in the 1950s and 1960s reported that they were having more sex than they wanted. Less sex isn't necessarily a sign of dissatisfaction. On the contrary, it looks like quality is improving even as quantity drops. Couples that are egalitarian tend to report higher rates of sexual satisfaction.[14]

But this leaves a vast number of people who aren't paired up and therefore don't have ways of seeking the touch that they want. While online dating should theoretically make it easier to meet and mate, we're so quick to reject people based on their profiles and photos that it's only the very attractive and successful who have the luck. In 2018, of the 1.6 billion swipes that people made on Tinder each day, there were only about 26 million matches, which is when both people are interested in the other. Of those, even fewer sent each other a message or planned to meet in person or ended up in a relationship. If the sexual revolution led to women feeling pressured and objectified, then the Internet has done the same to men. By judging each other's physical traits, we're preventing ourselves from finding warmth in real life.

Research shows that men struggle more than women due to lack of touch, but their feelings about it are ambivalent, meaning they have a higher desire for touch as well as an aversion to it. Their conflicted feelings appear to be associated with having an avoidant attachment style, which means they block themselves from receiving the affection they desire by being unavailable or unresponsive to their partners. This is likely because of social programming that tells them to avoid emotion, which makes them fear acting out on what they feel. Conversely, women report feeling more nurtured and joyful when cuddled than men do, perhaps because they're encouraged to embody the same qualities as they grow up, and they don't suffer as much when it's missing from their lives.[15]

We already know there's a good chance that denying our desire for human connection hurts our health and social fabric. In recent years the way we discuss loneliness has transformed from an unfortunate fact of life to a serious health epidemic with no end in sight. The issue has become so prominent that Britain assigned a "minister for loneliness."[16] The former U.S. surgeon general Vivek Murthy says the reduction in lifespan caused by long-term loneliness is equivalent to smoking fifteen cigarettes a day.[17] But when we think about solutions to our loneliness, we mostly consider mental and emotional ones, such as prescribing medication or developing networks of people to talk to. We don't think about the most visceral way that we use to communicate care, which is through touch.

No group better represents the collective consequence of a group of men having no one to touch than an Internet subculture called incels, short for "involuntarily celibate," men who report that women won't have sex with them. According to them, sex is relegated to the "Chads" or "Stacys," the genetically privileged—read: conventionally attractive—people of the world, a group they don't belong to. In their online discussions, some of them advocate for government policies that would require women to perform sex acts with them. Others endorse suicide or violence against more sexually successful people, venerating mass murderers who have made statements that have elements of the incel ideology.

These beliefs are obviously abhorrent. But when we discuss this group, we mostly talk about issues of gender and technology and race. We don't bring up as often their clear frustration and pain about having no human connection while at the same time having a deep desire for it. The incels' issues may seem unique to them, but they represent what happens to the human psyche because of touch hunger. Their loneliness transmutes into a toxic anger at the people they perceive as causing their problems—the beautiful people or those of a particular race. It's a hatred of the self, turned outward. The relationships they create from behind a screen can't replace the kind of connection they actually seek because what they're bonding over is their shared isolation.

What they actually desire is only possible in another's vicinity, as is the case of so many people stuck in a cycle of loneliness. We need to create more opportunities for closeness so our hopes aren't all tied up in having sex. If touch weren't such a precious resource—if we were able to get it from friends and not just romantic partners without being shamed—then it would cause us less grief. Another benefit would be that fewer of us would feel alone or agree to unfulfilling sex when all we want is affection. It might seem strange to readers of this book to think about asking for more closeness from our friends, but that's only because of where we've laid down our boundaries today. They're not permanent.

Avoiding touch and keeping to ourselves could make us feel more secure in the short term, but there's an even more serious and disparate distress that comes from having less of it over time. For centuries, we've become increasingly remote, which means even as we're trying to prevent unneeded friction between each other, we're also reinforcing old ideas about touch being dirty or a sign of sentimentality or lustfulness. We've become pretty rusty at meeting each other face to face and taking the risk of reaching out to someone new. I'm not the first to point out that it's no surprise that the more isolated we become as we sit behind our screens, the more polarized society is. We could benefit from reclaiming touch as a symbol of our shared humanity. The question is how we would get there.

In the late afternoon on November 9, 1965, the lights went out in New York. For ten hours, people were left to fumble around without televisions, and they were left twiddling their thumbs looking for something to occupy them. Nine months later, to the day, a splashy headline appeared in the *New York Times*: "Births Up 9 Months After the Blackout." As evidence, the article mentioned that Mount Sinai Hospital had seen quite an odd bump in births. Suddenly it became clear what people had spent their time doing during the blackout, robbed of peripheral entertainment, unable to find their method of birth control. Ever since the "blackout baby" story was birthed, any time a crisis occurs—a snow storm, a hurricane, an earthquake, or an act of violence—the media tends to predict a corresponding baby boom.[18]

Except the blackout babies were a myth. Often any perceivable increase in births is because people are paying more attention to the numbers, and otherwise meaningless blips are given purpose and meaning. There are exceptions. A notable one is the Oklahoma City bombing, after which Joseph Lee Rodgers, a University of Oklahoma psychologist, found there was a boom in that city after the event.[19] But in this case, it seemed the "boom" was a conscious decision. People in fear of losing their lives, as an insurance policy, made a choice to have another child. Rodgers says that going through a traumatic event often makes us adopt a more traditional mindset; it wasn't all driven by hormones. Still, the existence of the urban legend tells us something important about ourselves. We like the idea that, when we're totally immobilized and unable to go through our typical routines, we'll respond by clinging to each other again.

Adam Lippin, an entrepreneur from Montclair, New Jersey, recalls the panic of living during the AIDS crisis in New York in the 1980s. Over the phone, he tells me how tactile behaviors he witnessed then inspired him years later to start Cuddlist. He was in his twenties at the time and had recently come out as gay. He was already uncomfortable in his new identity, worried he would be ostracized at work if anyone found out. Lippin had observed that many of his straight friends had spent their adolescent years hanging out with girls and learning to initiate intimacy, so by the time they were adults, they seemed comfortable enough in their own skin. Lippin and other gay men were starting from scratch at an older age. He went from being alone to making up for lost time by becoming hypersexualized. Neither was satisfying to him.

Touch was already a nerve-racking prospect, but it was made even more complicated when he saw his peers getting sick and dying. Even though they would have found comfort in clinging to each other, even nonsexually, instead their attachment traumas became more evident. Either they avoided each other completely or threw themselves into dangerous and often unsatisfying sex or felt especially ambivalent about commitment even as they tried to maintain relationships. Their looming fear made them opt out of the kinds of social connections that could have been fulfilling to them. Some of them seemed to have no idea the kind of support they wanted or how to ask for it.

"I used to write in my diary that I want to have feeling based on actual human interaction rather than just mind talk," he says. "I know what it's like to be lonely."

As treatments for AIDS improved and those who had the disease were able to carry on without considering it a death sentence, Lippin slowly recovered from that traumatic time. He founded a successful chain of chicken wing restaurants and moved to New Jersey. He eventually found his way into the kind of committed relationship he wanted. But he never forgot what he and his friends had been through, and he observed that many years later several of them remained stuck, physically and emotionally, because of it. When he turned fifty, he wanted to do something more meaningful with his work, and his immediate thought was to help people gain greater body awareness.

Lippin thought about how others could benefit from being at the cuddle parties he'd gone to, get-togethers where people could exchange platonic touch. He enjoyed holding others and being held without any expectations of sex. Even though he was married, platonic touch reminded him that there was an opportunity for closeness all around him. As he did more research, he found out that there wasn't any formal training people could go through if they wanted to become professional cuddlers, so as someone with entrepreneurial experience, he set out to create a program for newcomers.

At first, Lippin received head-shaking and anxious responses to his idea. The idea of paying to cuddle seemed strange to many of his acquaintances. But he also knew that culture changes fast. Over his lifetime he's watched ideas go from kooky to normal, like sushi in the United States and onesies for adults. Once the training platform was established online at Cuddlist.com, he watched the community grow organically and people open up to professional cuddlers. More than a thousand people have gone through the training, and a little under a couple hundred are listed on the site as taking clients. It's the largest network of cuddlers in the world. For Lippin, it seemed like the kick in the pants that our culture needs to break free from our tactile constraints and open up to each other.

"I thought about all of those men I'd seen who had almost felt locked inside themselves. How do you get change to happen for them?" he says. "You put people in shocking situations. You can't solve the problem with the same conscience that created it."

I ask him if the clients ever feel that the touch they're receiving isn't authentic. He levels with me: for many of them, there is no other option. In an environment where touch is already scarce, if people don't have social capital or physical attractiveness or behave awkwardly around others, then they're going to be left out. We've created a set of cultural conditions—hectic work lives in big cities with plenty of one-bedroom apartments and restrictive social norms, for example—that makes it challenging for people to feel nurtured, and even if touch were more abundant, there are some who still wouldn't receive it. In a capitalistic culture, if we don't have access to something, the most obvious equalizer for those who can afford it is to pay for it. But that doesn't mean an exchange with another person lacks meaning.

"I hope the culture catches up and everyone has what we need," he says. "But that's not where we are right now. Is it really authentic, and is it real? The people I talk to say it's a very real experience."

Lippin is not a professional cuddler himself. So later I ask some of the professional cuddlers I've met who were trained through Cuddlist the same question based on their interactions with clients: Do people really feel the social the benefits of touch when they have to pay for it? Dawn interrogates the very basis of my question because it implies that affectionate touch that we receive in our personal relationships are somehow purer than those with a professional. But that's an unfair distinction because all relationships are transactional in some way. Sometimes this means we expect someone to text us back right away or to split the phone bill. Paying is just another such exchange. It's a way to receive touch without being beholden to this other set of expectations. "It's on a continuum, even though a lot of people think of it as odd," Dawn says.

While people are somewhat accepting of the drives behind sex work, we often view selling touch more harshly. Because physical contact without sex is considered to be more intimate, people don't like the idea of tainting it through capitalism. But the truth is these interactions don't matter any less or feel less pleasurable because they're part of a monetary exchange. They just don't lead to long-term monogamous relationships that we typically associate with these behaviors, and that pushes against the dominant scripts our society has around where it's appropriate to receive this kind of intimacy. It's exactly this thinking that keeps us in a touch-starved state. Dawn says that strangers become friends very quickly, and real relationships do form with clients even though the work is professional.

Shanks interjects to say that cuddlers aren't there to do whatever the client wants just because there's an exchange of money. A cuddler is trying to find a place of mutual comfort. One time he remembers going to a hotel to meet a man who contacted him online. They had a conversation in the lobby about what cuddling entails, including what the boundaries are. When they went to his room, though, it was clear he had something else in mind. Shanks says this client was rubbing against him and breathing hard. Shanks waited two minutes to see if he would calm down on his own, which was too long, before telling him he felt uncomfortable. After they changed positions, the heavy breathing continued, and Shanks tightened up. He should have left, but he gritted his teeth through the session. He was thankful when the man never reached out to him again because he wouldn't have ever seen him again.

It's a choice, in other words, to keep going back to work with the same people. He forms reciprocal relationships with his clients. Hiring a cuddler, to state to obvious, still means dealing with another person who has limits and needs of their own. Early in a session, Shanks usually tells his clients that hearing no from a cuddler is not a bad thing because it's a way for them to get more information. Most take that disclaimer well. But that brings up another worry about the future of touch, which is that when we have the option, we won't want another person at all. Instead, we'll want technology that can comply to our fantasies with no limits or expectations of its own. Some futurists believe that sex robots, and perhaps touch robots, will eventually become an acceptable outlet for people who feel isolated or whose partners are away. The appeal is clear; finally we'd have full control over the aspect of our lives that most confounds us.

But what happens when we can exert our egoic needs without any limits? If we avoid putting ourselves out there to another person, would it even be possible to receive the kind of care we so badly want? It could be that we need someone else, with their own preferences, to feel truly fulfilled. We don't yet know. A truth about human nature, though, is that once something—even if it's endless care from a robot—becomes freely available, then we move on to wanting something else. According to the *New York Times*, now that the middle class can afford technology, the rich want to spend their money on experiences with people instead. They're embracing tech-free schools and expensive spa treatments to escape their digital lives. If robotic touch becomes an accessible norm, we could be driven right back into each other's arms.[20]

In her 2017 novel *Touch*, Courtney Maum creates a fictional world in which people are opting for personal touch again. The main character, Sloane, a trend forecaster with a "spongy sensitivity to her environment" who is estranged from her family and is in a sexless partnership, is hired by a company to figure out how to launch products for a growing population of voluntarily child-free people. But throughout her assignment, she starts to see that what the future actually holds is a complete reversal. People are craving closeness and family. Sloane pictures a reversal to "in-personism." She proposes monitoring by doctors to determine how much physical contact people have gotten and salons where people can pay for hugs and apps to rent children for the day.[21]

Whichever direction we move—toward the human-centered touch that cuddlers offer or the robotic version—the commodification of touch can seem like part of a bleak future. But really it's been an ongoing trend since the industrial age. As we've grown further apart and vision has continued to dominate our sensory landscape, our consumer culture has stepped in as a replacement, though it remains unclear if it's a real solution to our touch deprivation. Cuddlers are just a small part of how we're turning to the market for touch that we aren't receiving elsewhere.

7

Slick

How Companies Sell Us Touch

Emiel DenHartog, associate director of the Textile Protection and Comfort Center at the North Carolina State University School of Textiles, walks me through its entrance, where there's a large display of clothing—a hockey uniform, running shorts, and hospital scrubs—that has been examined here recently. Everything from ordinary clothing such as underwear to high performance gear is sent to the center to be evaluated for how the clothing feels to the touch. Certain fabrics, like those used by firefighters and industrial workers, need to serve special functions such as handling heat or surviving extreme wear and tear while also moving with ease, so it's especially important to look at their comfort level closely.

One of the lab's workers takes a square swatch of fabric, which she says is going to be used for parachutes, and mounts it on a machine that holds it flat. A loop-shaped probe brushes lightly across the top to measure the force required for the probe to move over the fabric's surface to check its textural properties. This is just one of the tests that go into what's called the Kawabata evaluation system which relates people's perceptions of a fabric's "hand" to its mechanical properties. This system, which was invented by a chemist named Takeo Kawabata in Kyoto, Japan, in the early 1980s, measures surface texture but also tensile (ability to be stretched), shearing (ability to be draped), flexibility, and compression. A separate set of tests evaluates the temperature qualities of a textile, such as how fast it allows heat or water to evaporate.

The machines spit out a spreadsheet full of measurements, which are incredibly valuable to brands. Industry is constantly trying to improve materials, such as the moisture-wicking quality of yoga pants or the cooling effect of a bedsheet, or adding an element like a sun-protecting nano-coating to a T-shirt while keeping its texture the same. This is how fabrics are assessed. DenHartog's work is an example of the field of sensory evaluation, the science of how our products are made to look, sound, smell, taste, and feel nicer, and it is one way that society has commodified the senses in recent decades. This research is not unique to textile manufacturers. For a variety of products, feeling is something that can be engineered and promoted. Just as synthetic fabrics are judged on their wearability, chips are rated on their crispiness, and shampoo is designed to cool and tingle the scalp.

But how to evaluate feeling is incredibly complicated, more so than the other senses. DenHartog admits it's not possible to say much about an item's comfort based on a readout from a machine. A person—whose evaluation could be totally different from the test results—still needs to wear it. That's because preference depends on factors that can't yet be measured, like how a garment moves against each person's body. And our decisions aren't guided by physics alone. The texture we expect from our jeans is different from that of our pajamas, and we have decided what we want from each after years of experience. We're more confident, and therefore more comfortable, when what we feel matches our expectations, which are highly subjective. It would be nice to eventually let the numbers do all the work, but it will be a very long time before we crack the code of consumer choice, which is a mix of physiology, prior experience, and personal predilection.

This raises the question: Then what's the point of the data? DenHartog himself says he doesn't think too deeply about the feeling of his clothes or his bedsheets when he's out shopping. For most everyday items we use it's not complicated to make something that's pleasant to touch. The answer is that it's about marketing. DenHartog and I get to talking about yoga pants again. I recently bought an outrageously expensive pair that claim to wick away moisture as well as a bird's wing, or something like that. DenHartog's face tells me that I'm a huge sucker. My pants might absorb slightly less sweat than another fabric, but he says that's not the reason I bought them. Rather, marketers have been successful at selling the concept of moisture-wicking fabrics to customers, and now we're all expected to wear gear made with this fabric if we want to feel stylish in public. If I feel better in the pants, it probably has more to do with being part of the in-crowd than aridity.

"Let's just say comfort is probably one small factor for people who do normal workouts," he says. "If social expectations changed, and we all wanted to show off how much we sweat after exercise, we would go back to cotton. That's something I think about a lot."

In other words, even though on the surface it might look like what we care about is touch, we could actually be driven more by vision—a desire to blend in with others. The entire field of sensory evaluation would suggest that we're finally placing more importance on touch and perhaps other unappreciated senses, but it's not so simple. There are numerous ways that the practices used to optimize all of the senses continue to reflect our bias about the preeminence of vision. As I tour the range of products and services we have today that are made to appeal to touch, I'm reminded repeatedly of how our sensory hierarchy remains intact.

Some of the most sophisticated work in sensory evaluation is done at the Nestlé Research Center in the town of Vers-chez-les-Blanc, Switzerland, set in a landscape of rolling hills dotted with neat little homes with red roofs, grazing sheep, and twirling dandelions. Fronted by a quaint cobblestone road and a manicured garden, the center blends right into this scenery. It's a Disneyland-like place, meant to look like the home of an artisanal food producer, even though this is the origin of Hot Pockets and Goobers, and it has behemoth power over what the world eats. I drop in to learn about research on texture, or what I like to refer to as "mouth touch."

Nathalie Martin, the group leader of the Behavior and Perception Group; her colleague Benjamin Le Révérend; and some other staff members sit with me around a conference table. Martin has sun-kissed skin and is wearing an airy silk blouse. Le Révérend is boyish, and he looks owl-like in his Harry Potter glasses and has on a stiff shirt—which makes him appear comparatively geeky. They each don their part well; she represents the soft sciences, and he the hard. Martin studies perceptions of pleasure in people, and Le Révérend uses machines to come up with numbers that support her findings. Martin says texture is particularly difficult to study compared with the other qualities of food because it involves so many different assessments by the tongue, mouth, and brain. If you're trying to make something sweeter, then you can just add more sugar or sweetener. Texture isn't so linear. The qualities we enjoy, such as creaminess, tenderness, and crumbliness, have clear meanings to us, but they can't be manipulated so easily.

"There is a beauty and complexity to these interactions," Martin says. "On the one hand the reported experience is quite simple. But to fully understand how people come to that interpretation, you need to know what happens in the physical and chemical interactions in the mouth with the product."

To help me understand, Martin walks me through some of her studies at the company on the subject of crossmodal associations, or how the senses influence each other. Early in her time at the company, she was asked how to make yogurt feel thicker without changing its consistency. Actually making the yogurt more viscous would mean altering the amount of pressure needed to move it through pipes in a plant, which could introduce all kinds of unforeseen complications. Tricking consumer's minds seemed like a much easier bet. Martin explains that people are highly selective about the products they use, but they don't know why they make the decisions they do, even when they do offer up an explanation. Her job is to figure that out.

Martin had read prior research that showed adding a thickening agent to yogurt made some people perceive it as less flavorful. But the thickener wasn't affecting the flavor compounds or trapping them in the bowl, so there was no physical reason why. It seemed this effect was an illusion created by tasters' minds. Prior culinary experiences had taught them to feel thickness and associate it with added fat, and we all know that fat mellows out extremes of sweet and spice. So slightly higher viscosity automatically made them taste less. She believed that the opposite might be true as well—that if she made the flavor of the yogurt blander, then tasters might just assume it contained more fat.

To test her theory, the group made a series of yogurts with two different thicknesses and added various olfactory compounds.[1] The reason is that, along with taste, aroma accounts for a lot of what we consider to be a food's flavor. They presented the yogurts to sixteen lab workers. They found that when the yogurt contained just one chemical, such as ethyl butyrate, which gives off a flat, pallid strawberry scent instead of a more robust natural one involving several more ingredients, testers rated the yogurt as thicker. Her hypothesis was right. The type of aroma also affected perceptions of texture. Yogurt with butter and coconut notes was perceived as thicker, likely because we associate these flavors with heavy foods. The research showed that both texture and flavor create our impression of unctuousness.

After the yogurt study, Martin investigated another crossmodal association. This time she was asked to improve a chocolate bar without changing its recipe. She began with the idea that the shape could affect its feeling. A dome-shaped piece would expose people's tongues to a larger surface area of melting chocolate, which might make them rate it as creamier. She prepared pieces of chocolate of roughly the same size in ten different shapes, including a rectangle, a triangle, a sail, a bird's wing, an oval, a trapezoid, and an egg. She and her team made a few trial batches with each mold and had forty-five trained tasters try them.[2]

Her hypothesis panned out, but with an unforeseen consequence. The rounded shape did feel like it was melting better to tasters. But they also rated it as having less chocolaty flavor. The increased surface area left less room in the mouth for the aromas to swirl around and float up to the nose. The group worked out a compromise that could enhance both the flavor and feel of the chocolate. They fashioned each piece into a trapezoidal shape and increased its surface area for melting with a convex curve, still leaving plenty of room for air in the mouth. The resulting candy bar was sold in Switzerland.

The texture research at Nestlé has ranged from the way the texture and enjoyment of crunchy cereal wanes while being masticated into a sludgy bolus, to how matte paper-like packaging can make the product inside seem more natural to consumers than a plasticky finish, to how the length of time spent chewing on a food item affects satiety. Most of Martin's work is done with panels of either regular people or professionals trained to detect qualities of food. But these groups can be expensive to put together. And human beings, regardless of their skill level, are not all that reliable. They're not that great at telling the difference between, say, crunchy and crispy or a creamy taste and a creamy feeling. And our mouths are attuned to certain types of sensation while ignoring others. And how we want our food to feel alters how we experience it.

That's why it's helpful to convert all these descriptions into a set of physical and chemical properties, which is where Le Révérend, comes in. He uses some of the same tools as speech pathologists on his test subjects—electrodes that he places on their temples, the bones behind the ears, their jaws, and the opening of their mouths—to record the electric activity produced by their muscles when teeth gnash together. After the chocolate study, he made some members of the research center wear another one of his instruments, special goggles with a nose attachment that could measure how much aroma was released as the various shapes melted on their tongues. What he found confirmed Martin's results; the rounder shapes did release less of the chocolate's scent. It was one of the rare times it was possible to turn a tactile experience into a measurable figure.

"It's a big challenge to understand the mouth as a machine, what kind of sensors it's equipped with and the relationship between perceived texture and physical structure. It's fantastic because there's a very fundamental physiology to be understood which is still very new," Le Révérend tells me.

The sensory evaluation of food has a fairly short history. Beer brewers in the 1930s were the first to start examining the sensory qualities of their beverages, Martin explains. They used categories such as fullness, thickness, astringency, and sparkle to help consumers choose beverages that appealed to them. Such terms were far more specific and informative than those adopted by the wine industry—the notes of cherry or charcoal—which are poetic but not so precise. But the field really took off in the period following World War II, when the food industry applied this approach to its own purposes. Special methods that had been used to preserve food for troops were now readily available at the grocery store, and companies tried to make sure they could be as appealing and consistent in their makeup as possible.

The exact parameters of texture were the most important factor in ensuring that large-scale food production ran smoothly because food was made in plants where viscous sludge had to flow smoothly through giant pipes. At the same time, this consistency needed to transform into something that people wanted to eat. Because the intention was to get consumers to escape their rational minds and to convince them to buy what they didn't need, much of the focus was on the so-called lowest senses—touch, smell, and taste.[3] This meant a food item's physical makeup must correlate with people's psychological experience of eating it. The work, then, was just a matter of taking subjective judgments—chewiness, tenderness, crunchiness, crispness, and all the other "ness" words—and transforming them in to objective measurements.

Science, rather than intuition, was being used for the first time to feed the masses, so there were new protocols needed for this research. Labs based their methods on trying to find objective measures for sensation as possible. The very definition of objectivity only came to be understood the way it is now alongside technological developments and rapid industrialization of the nineteenth century.[4] For the first time, the widespread use of photography allowed scientists to slow down and capture moments that occurred too quickly and were so minuscule that the human eye couldn't observe them. A new ethos arose around scientists aiming to be as precise and unobtrusive as their machines, which meant that even in fields that didn't use images, new practices evolved to prevent scientists from adding their personal interpretations.

Food companies used mechanical mouths to measure texture to create a separation between themselves and what they were studying, hoping to further the goal of objectivity. One such device was the strain gage denture tenderometer, a machine that looks like a set of dentures attached to two moving aluminum jaws. The companies believed an instrument would be more reliable than a human mouth, but they found instead that there were limitations to what it could do. Eating is a complex, dynamic process, and our mouths alter what we consume, moving it around from side to side, chewing it with various parts of the teeth while adding saliva to the mix. No machine has ever been sophisticated enough to replicate all of these functions. Nor can machines make the kinds of subjective judgments that the human brain does when eating. This is why many scientists working in the field of sensory evaluation now have reverted back to using humans for their tests and then confirming the results with machines. But this approach poses other challenges.

Le Révérend says that often there isn't an assessable cause for why tasters prefer one texture to another. Perceptions are usually not based on facts alone. He leads me to the company's cafeteria, which has the fanciest cappuccino maker I've ever seen. He tells me half-jokingly that it cost as much as a small car. As I sip from my small paper cup, he describes what is happening in my mouth. Coffee on its own has such a strong smell because its aroma compounds don't like being mixed in water, so they escape out of the cup. But these compounds do like fat, so adding milk forces more of them to stay inside the cup, mellowing out the flavor. There are also some new milky notes, a kind of "fat taste" that's added. He poses the question to me: "To make a low-fat version of this cup that tastes as good, how would you go about it?"

If the yogurt study has taught me anything, I know the solution might be to find a way to lower the aroma without adding fat, to fool drinkers into thinking they're drinking something that's milkier than it is. He tells me the company's scientists had suspected so as well. To test whether it would work, Le Révérend made people drink coffee diluted with various amounts of milk with their noses plugged. But something was off. The tongue's texture sensors were far and away more sensitive than a machine's, so people could tell the difference in thickness between the fluids whereas a machine could not. The study shows that for every food item, the rules around touch perception seem entirely different.

Whenever we touch, we assess both physical *and* affective components of feeling, but scientists don't like it when our messy human judgments get involved. They choose to ignore the subjective factors that lead us to make our choices and instead try to find an underlying logic them. This is a useful method for coming up with workable experiments. But in the end, each study reveals only one small truth, and there are always exceptions. It seems that by ignoring the emotional reasons we prefer some textures over others, scientists are eliminating a key aspect of what they're trying to understand. Although they'd like to find a quantitative relationship between a specific stimulus and a human response, there's no real evidence that's how we sense at all.

While sensory evaluation started with food, it has quickly spread to beauty products, clothing, and even our cars. Almost every pleasant texture we come up against is created by scientists thinking about the subtle interactions between the body and brain, and in all these industries there exist similar puzzles. What they're all learning is that objectivity has its limits. The word itself is directly related to the eye and rooted in the belief that the distance senses, particularly vision, provide a superior way of knowing. Touch can only be fully understood only up close, subjectively, but so far we don't have a widely used approach to scientific research that allows for that.

The process of sensory science for large industries is not the only way our modern culture seeks to celebrate touch while simultaneously sidelining it. The entire design of contemporary life makes our senses seem separate from each other, as if we need unique spaces to experience each of them. We listen to music at concert venues. We eat at restaurants. We receive touch at massage studios and relish our bodies' movement at gyms. This has occurred as our full sensory experience has become limited day to day. As we spend more time indoors, sitting at computers, hardly interacting with anyone, where we come back to feeling is in a space precisely fabricated for it.[5]

Nowhere is this development more clearly felt than in medicine. Centuries ago medical practitioners believed that the types of sensation we were exposed to, including touch, were the cause of illness, which is why the cure often also involved the senses. In ancient Greece, for example, four key tactile properties in the body—heat, cold, wetness, and dryness—needed to stay in balance to maintain health. The odor of a swamp nearby could lead to too much dampness, and the way to relieve it was through the senses too, through diet, pleasing scents, or nurturing touch. Honey was considered warming, and the cooling scent of rose was recommended in summer. Certain colors were believed to affect the elements of the body, which is why people were instructed to wear particular gemstones and minerals. Similar practices existed all over the world.

"Here are the things that will relieve one's exhaustion: the showers and cooling breezes of sprinklers, and the gentle wafting of fans made of palm leaves and broad lotus leaves; garlands of camphor and jasmine; pearls with yellow sandalwood; the delightful low chattering of baby hill mynas and parakeets," writes Dominik Wujastyk in the collection *The Roots of Ayurveda: Selections from Sanskrit Medical Writings,* published in 1998 about the traditional Indian health system.[6]

Just as the senses were central to the understanding of illness, they were also the primary way early physicians made diagnoses. These practitioners looked at a patient's physical appearance, including their coloration, often from various angles. They smelled their skin and listened to the sound of their breath. They tasted urine. Through touch, they could sense the tightness and looseness of various organs. Pulse-taking required extremely sensitive hands. Four fingers were used to take account of the speed, strength, regularity, and rhythm of the heart's activity. Massage was a natural way to provide care. Touch was, to many healing cultures, the most important sense of all.

A sensory shift began to occur in the sixteenth and seventeenth centuries because of the influence of Protestantism. Premodern medicine was still based on many of the same theories of illness as in the past, but it started to feel different. Pleasurable treatments, such as massage or eating sugar, which was once though healthful because it was so easy to digest, fell out of favor. Religious teachings said that sensory pleasures could corrupt morals, so it made no sense that they could be good for health. There was a belief that medicine had to be a tough pill to swallow. People were also becoming less comfortable with physical contact, so examination techniques involving touch fell out of favor. Patients preferred to self-report their cases instead of having a doctor get close to them.

By the nineteenth century, microscopes, thermometers, and stethoscopes were adopted by doctors. However, this wasn't because of their utility. Thermometers, for instance, had been available since the seventeenth century, but they weren't accepted immediately by doctors, who were convinced their God-given senses were superior. It was cultural change, instead, that led to their eventual acceptance; doctors and patients were coming to prefer more personal space, which these tools made possible. As a new set of practices grew around the need to survey the body from a distance, sight became the dominant sense in medicine, replacing touch. Because sight was already the primary sense in most of science, the lack of touch became a sign of professionalization. In this new era, doctors had to remind each other, as they still do, that there's an important ritual that occurs when doctors touch patients.

"What is the secret of success? To inspire confidence. What is confidence? It is a magic gift granted by birth-right to one man and denied to another. The doctor who possesses this gift can almost raise the dead," writes Axel Munthe, a doctor in the early 1900s.[7]

Today, visiting the doctor is markedly nonsensual. There's very little character in the setting of a clinic or hospital. The walls are white and flat. The smell is antiseptic. The food is notoriously flavorless. The goal is to offend a patients' sensitivities as little as possible. Most of the time there is spent waiting—for the physician to be ready, to have blood work done, to get scans taken, and for the information to come through. Even nurses, whose job is to substitute for a doctor's senses, for the most part only touch patients in perfunctory ways, to adjust a pillow or to prevent a bedsore. Sensory forms of healing—pleasant flavors, sounds, and feelings—are completely missing. While we've adjusted to these norms, there's evidence that we still have a desire for older, more primal forms of caregiving.

As we subtract a sense from daily life, it tends to pop up in a new consumer trend, and that's exactly what we can observe in the popularity of alternative medicine. People achieve a sense of immediate restoration when they stimulate their skin with a dry brush or soak in a thermal bath or use therapies such as qigong, ayurvedic medicine, and acupuncture, which emphasize the physical aspects of healing. At Ayahuasca retreats, people experience the senses merging together, a distraction from their pain. While many professionals will dismiss these practices as quackery, their popularity shows us how important the aesthetics of healing remain to us. Western medicine might be better at providing a diagnosis and treatment, but it doesn't lead to a full-bodied feeling of wellness.

There are endless stories of skeptics finding relief this way, so I'll share just one. Meg Freeland, a bubbly and articulate physical therapist based in Brooklyn, tells me about her neck pain that started all of a sudden when she was a student in 2002. The pain radiated from under her shoulder blade up to the base of her skull. An X-ray confirmed that one of the disks in her upper spinal column was bulging, but doctors weren't sure of the cause, and there was nothing that could be done about it. They told her to rest when she needed to and avoid putting strain on the area. Her busy life, however, made it impossible to slow down much, and her struggle didn't end. She became used to living with the on-and-off pain, and the episodes became more frequent and intense. They hit every few weeks and grounded her for days.

Eventually, after she graduated, it became difficult for her to make it to her regular shifts as a physical therapist at a prominent hospital. Sometimes the pain was so excruciating that she dreaded even breathing. She saw a neurologist and a physical therapist, hoping they could help her out. But they couldn't. She had local anesthesia injected into the tense portions of her neck, but the pain kept getting worse. She went for massages, and although they didn't cause her any more problems, they didn't result in any improvement. Seeing how much she was suffering, one of her coworkers, who was receiving training in an immensely popular but unproven treatment called craniosacral therapy, which involves light touch for pain management, suggested she come see him.

Freeland knew about craniosacral therapy; she simply didn't believe in it. The practice was developed in the early 1900s by William Garner Sutherland, then a senior student of osteopathy. During his schooling, he was taught that the adult cranial bones don't move because their joints are fused together. But when Sutherland felt for motion in these bones in patients, he found that they did oscillate slightly, in a wave pattern. He believed that the movement he felt was the breathing of the tissues, or a life force. And he developed the theory that by feeling for the rhythm, he could sense restrictions in tissues and help relieve them. Many doctors write him off completely since no one has been able to verify the kind of movement Sutherland claimed. Still, the therapy he proposed took off. Today, Sutherland's approach is used not just on the cranium but also the sacrum and tailbone.

During one bout of serious pain, Freeland was ready to try anything. She went in for her first craniosacral appointment in so much agony that it took an entire half hour to make her way to lying flat on the table. Then her therapist had her lift her hips and rest her bottom on his hand while he made very slight adjustments. The therapist asked her to let him know if she felt anything. She told him at one point that her left ear got hot and there was tingling at the top. Later there was what seemed like a rush of blood in her arm. After an hour and a half of treatment, her pain was no longer debilitating. It wasn't gone, but it had reduced significantly to a level she could manage. She started going regularly, every couple weeks. And over months, the pain dramatically reduced.

"It was literally the only thing that worked," she says.

It's hard to find explanations for stories like hers. Studies on alternative therapies are limited because treatments that require the individual attention of a practitioner are harder to provide in a uniform way for the purpose of research and can't generate as much revenue as a pill, so there's less financial incentive to do them. Also, the reasons they work are likely to be extremely diffuse. While such treatments may work physically in the ways that practitioners claim, another important part of their benefit is how they affect our larger, amorphous understanding of our health, which includes our mood. It feels good to receive the patient, unhurried hands of a healer telling us we're going to be okay. It makes us leave a session feeling immediately better, which can spiral out into even larger improvements outside of the office.

It's no accident that when patients have given up on every other treatment, when they feel they've lost hope, they report finally finding some relief through the hands of an alternative healer. Most of us would appreciate having such a person paying careful attention to our overall well-being, but only a small fragment of us can pay for these adjunct treatments. The sad fact is that our simplest forms of nurturing the sick—a hand on a shoulder, the sympathetic presence of another person—are out of reach for a vast majority. There could be a lesson here for medicine about the value of the interpersonal bond between a patient and doctor and the need to return to some of our oldest forms of healing, but the current medical system doesn't free up doctors to practice this softer side of medicine, and there's no sign this is changing.

One New Year's Eve a friend and I decide to skip the parties and take part in a midnight yoga class. We're in a room packed mat to mat with people with the same idea, clad in multicolored Lycra. The lights are dim, and for an hour and a half we contort ourselves, gently, into a variety of poses. It's a cold night in San Francisco, but inside, even without any heat, the air temperature warms up amid so many bodies in motion. As our muscles shake through a long, painful stretch, one of our yoga instructors reminds us that our minds are wonderful at getting work done and sorting out problems, but when we can't shut them off, they can become our prisons. By tuning into the feelings of our bodies, including embracing this pain, we can live in the present moment, outside of those racing thoughts. She encourages us to let ourselves feel it as we sink further into the stretch. It's the New Age version of exactly what science tells us—that we are susceptible to the constant disruption that come from outside of us, and returning to our sense of touch can ground us within our physical realities and emotions.

As midnight approaches, she asks us all to come close to the center of the room and place our hands on the backs of someone else until we're linked together in the shape of an imperfect mandala. We chant "Oooooooom" over and over, for half an hour, taking long breaths in between. The room gets even more humid, with so many sweating bodies pressed close to each other. The vibration of our larynxes is creating what feels like an orb of energy around us, like it's shaking each air particle. When finally we stop and have a moment of silence, the vibration ceases and calm is restored, like a collective sigh that I can sense deep within me. I turn to my friend on our way out and ask her, "Did you feel that?" She nods. I feel restored and ready for a new year, but almost immediately upon leaving, I'm not sure how to think about our celebration.

In one light, it's a night of beautiful symbolism and group ritual. What we've done here goes beyond fad. Exertion is the simplest, least-conceptual way of short-circuiting our ruminating brains. And group exercise is a way to tap into feelings of unity with each other, especially in a culture where religious practices have fallen out of favor. By moving our bodies in unison, we feel like we're creating a shared energy in the room. Meditation, physical discipline, and fasting have long been a part of our world's religions, and wellness has adopted them for similar reasons. But through another lens I'm the caricature of a millennial. I've paid eighty dollars and much more than that in yoga gear for this privilege of appropriating an ancient spiritual practice. There's a well-worn argument here about colonization and oppression. But there's also something to be said about how the way we use yoga reinforces old racial beliefs about touch.

The truth is that we don't need to look at some older, purer culture outside of our own to become more in tune with our bodies. There are ancient Western practices that we could also embrace. But by looking elsewhere we continue to hold a superior position as a vision-centered culture while claiming that other, more primitive people are more acquainted with the lowest of senses. While we say we venerate cultures that celebrate the body, it is telling that we turn to them in an effort to shut our minds—our most precious tools—off. This is just a continuation of racial biases we've seen in previous iterations, and it's as true of yoga as it is of spas that advertise Chinese *gua sha*, a vigorous scrubdown of the skin to increase circulation, or sauna treatments that mimic Native American sweating lodges. Our feelings about touch are just as conflicted as they've always been.[8]

Who is allowed these sensory indulgences is also based on deep prejudices. Social expectations tell us that only some people are allowed to concern themselves with their own comfort; some are expected to live the life of the mind and others are meant to endure more roughness in life. The world continues to feel different to us depending on our gender, race, and social standing. White women are the most likely to benefit from the litany of tactile commodities available to us. Meanwhile, men are rewarded for toughening themselves. People from marginalized groups might get the feeling that these tactile services are not for them. We use commerce in the same way we've always used interpersonal touch to maintain a stratified society.

"Social identities remain but as one is turned into a consumer, they are increasingly shaped and conditioned by patterns of consumption. We identify our real selves by the choices we make from the images, fashions, and lifestyles available in the market, and these in turn become the vehicles by which we perceive others and they us," writes Joseph E. Davis, University of Virginia sociologist in his essay "The Commodification of the Self."[9]

A friend tells me that I should try out the most outrageous spa treatment they've heard of—a flotation tank—where I can lose all my senses, and I immediately know I have to do it. A few weeks later, I'm receiving instructions before heading in. The owner of the facility asks me what brought me here. Book research, I say, but I also want to know what it's like to be free of my senses. He tells me about his own life-changing experience, during which he experienced what he termed as "ego death." Without any sensation, he was able to escape his body, finally understanding what it was like to be only a soul. Ever since, he's been less afraid of his own mortality. "In the West, we're so afraid of death," he says. "We don't live with it the way Eastern cultures do." I tell him he'd be surprised that much of the Eastern world isn't too thrilled about death either.

Soon I'm in a dark room, lying at the surface of a tub of body-temperature water filled with Epsom salts, which is supposed to give me the impression that I'm suspended in air. I've been told that I'm supposed to lose my connection to my body completely. But the opposite happens; I'm hyperaware of it. I can still feel the wetness on my skin and how the sound of my breath is magnified by the water. I am so attuned to these small details that they occupy all my thoughts. I change my focus and think instead about what I can't sense. I can't tell whether I'm tensing my neck or letting go completely, whether it is my muscles or the salt water that is holding me up. I'm ruminating.

I try again. I let myself simply be. After soaking for a little longer, I can't tell where my fingertips end and the water begins. I'm being supported effortlessly, so I feel light and unburdened. Unoccupied with every faint sound and feeling, a new set of thoughts become louder to me. I start to solve all the problems that I've been piling up for weeks and months—how to redo that paragraph I've been working on, how long I'm going to stick around in New York, what I hope to do with the next decade of my life. Without any knowledge of the passing of time or the feelings of my body, my mind races, and I can't believe when it's time for me to wrap up. I imagine this is what advanced meditators feel, that sensation of levitating out of one's skin and watching oneself from up above.

Once my time's up and I'm back outside, it doesn't take long—minutes, maybe—to feel anxiety creeping up on me again. My phone buzzes, and I remember that the day has flown by, and I haven't gotten anything done. I make a mental list of everything I have left to do. I wonder whether the float tank was worth it. Was it meant to be enjoyed in the moment, or did I expect to carry its effects for weeks after? This is the flaw of our perpetuation of the belief that the senses are completely distinct from each other and that we need special services to experience them. When we relegate our enjoyment of the tactile to a corner of our lives, we have to leave it behind when we're done. The crucial question is whether there's an alternative to all of these complicated practices, which is to spend time moving and making physical things and embracing embodied forms of human interaction.

I own a fake fur coat. It's somewhere between burgundy and hot pink, and it's so thin that it can't keep out any cold. Wearing it feels at once luxurious and wicked, and it's eminently clear to me why. It's a costume that represents how I've been taught to think about my sense of touch—the appeal of comfort and luxury but also the presumption of vanity. Fur has historically been accessible only to the rich, and for rich women, who didn't have any real authority compared to their husbands, it was a way to display their proximity to power. We see female dominance through a distorted lens, so it's no surprise that wearing fur has become a symbol of our kinkiest and most vilified characters, from Wanda in *Venus in Furs* to Cruella de Vil. Fur was vilified before campaigns about animal cruelty made it taboo.

I'd never studied the cultural history of fur, but nonetheless I somehow know exactly how I should feel when I have my coat on. I've absorbed messaging from images of people who wear it, how we talk about it, and how it's sold to me.[10] In that same way, I imagine that everything I touch is embedded with so many values and beliefs about how much I should care about the comfort of my body and when I should feel guilty for overindulgence. We're hardly aware of how much our tactile past influences us, and yet the current marginalization of touch is an outgrowth of what we've unconsciously been absorbing for years. What we touch constantly shapes us, even when we don't notice it.

I share this example because it shows the processes that keep us stuck in an old mindset, even as we try to incorporate more touch into our lives. Some theorists would say we are in the midst of a sensory evolution. The desire to unplug, the excesses of the market, and the rise of the professional cuddler are all driven by these same desires—to make tangible our place amid our surroundings. It's easy to scoff at all these leanings as hipsterish or New Age. But instead of being cynical about it, it is useful to look at them as a critique of where we are. Our culture and our technology are conspiring to pull us out of our bodies, and people are recognizing their yawning need for touch that isn't being satisfied. They're trying to rebalance our senses. But we can't shake the old scripts.

The variety of ways we have to appeal to touch only continues to grow. Ever more consumer products and services have popped up to fill our sensory void. We have sheets that caress us, shirts designed to hug us, and massage chairs to soothe us. Our weighted blankets and fidget spinners give us diversions to soothe ourselves. We have robotic pets that can rumble and purr and don't require work or cause a mess. Ever since industry began its efforts at sensory marketing, our consumer goods have felt more recognizable to us than the sensations we experience in nature. And yet these feelings aren't necessarily what drive us. We like the idea that they do, but we may just be responding to marketing. We no longer go to stores, where we paw at these products in a tactile showroom. Instead, we're purchasing them online, based on visuals.

In the past decade, there's been a huge upsurge of body-centered practices. Yoga and tai chi are all about tuning into feeling the stretch of the skin and muscles and the slow movement of breath in and out of the lips. These exercises ask us to get out of our heads and gain a close, reflective awareness of what's happening to our bodies. The spa industry is worth about $17.5 billion a year, according to a 2017 survey. We're creating new and ever more elaborate ways to feel the kind of connection to our bodies that's missing from most of our interactions. But the origins of these practices and who is selling and buying them tells us even more about how we continue to think about touch.

More of us are turning to the kinds of activities our grandparents did for more practical reasons, such as canning, woodworking, gardening, knitting, and brewing beer. The entire Maker Movement is based on the ethos of good old American self-reliance as well as the more modern availability of online learning and modern design. But it's also about a return to touch. The resurgence of old-fashioned technology such as vinyl LPs and typewriters represent a desire to reinhabit a physical space. As more of our products are made by machines, there's a growing cachet that comes from buying handmade and eating at restaurants that are small, locally sourced, and less mechanized. We want to surround ourselves with the kinds of products that another person could have made with their hands. But these are all just small replacements for the lack of hands-on experience in the dross of our daily lives, which don't appear to be returning.

Even though we're seeing minor modifications, it turns out that many of our ideas about the senses remain. Our plethora of products and services make it appear that we're finally seeing value in our lowest sense and correcting for the missteps of the past, but through another lens we're continuing to marginalize touch in new ways. Whether going to a wellness retreat or a float-tank session or buying individual products based on feel, we're just temporarily distracting ourselves from what's missing in our continuous reality. However, there's still one more area we haven't looked at yet where industry is trying to restore touch, and it could be the most impactful.

8

Haptics

Bringing Touch to Our Technology

Ingo Koehler, the leader of the haptics group at Volkswagen, is a touch artist. At least that's what he sounds like to me as he describes the process of coming up with the feelings of the buttons and dials on all the company's cars. He and the team he leads tackle their jobs as storytellers, with hopes that the act of adjusting the air conditioner or changing the radio station will convey something about the brand the way the car's shape and logo do. He has me compare a few of them from the various brands owned by the Volkswagen Group, which are splayed out on a table.

"The haptics are really different," he says. "Audi is more about the click, the acoustic. I think it's a little more technical sounding. The Porsche is sporty. It's harder and shorter. It takes slightly more effort to push. The Volkswagen brand is meant to be comfortable. The haptic uses less sound and has more feeling."

He's right. It's amazing how much they vary. As I play around with them, each control starts to feel so distinct to me that I can guess them without looking. He says he arrived at these haptics by surveying his colleagues. He made several samples with different materials that required varying amounts of pressure to use. Once they coalesced around a choice, he sent the instructions for how to reproduce the feeling, including a force curve describing the pressure needed to use them, to the suppliers who manufacture them. I ask if he has a favorite in the collection.

"No, you can't really say what the best is; this is the best haptic, this is the second best haptic, no," says Koehler. "We found you have a good haptic if the feeling matches your expectations of the brand."

He believes it is subtle improvements—to the haptics, for example—that make Volkswagen Group's cars stand out these days, now that all automakers have reached certain standards for performance and safety. He can never be sure that the way a particular button yields to a customer's touch is the reason someone chooses one automobile over another. But he got a pretty good sense of how impactful it was in 2001, the first year the company debuted a car whose buttons and dials were all built to be uniform. Until then, the contractors who made the various parts weren't given any haptic instructions to follow, so they all felt slightly different.

"Golf 7 was the first car with the new haptics," Koehler says. "Everyone thought that it was a real jump in quality. Newspapers and car magazines were saying it. We didn't tell anyone about the haptics, but we think our team had a lot to do with that because the materials on the seat and the dashboard didn't change. What changed was the haptics. The parts all matched. They were harmonized. People are good at telling differences, but they won't say this button is 3 Newtons and that is 4. They will say if it feels sloppy or it feels nice, and this felt better to them."

After years of refining these controls for each new model, today his group is facing another challenge. Many brands are switching over to control panels with touch screens, a look that consumers have come to expect from their electronics. Where there used to be three-dimensional parts to manipulate, there is now just a flat screen, which means most of the haptics are gone. This is a major safety hazard because, without that feedback, we don't know if we're pushing the right buttons without glancing down. The team is at work finding new ways to produce feeling in the touch screen, to get drivers to keep their eyes on the road.

Koehler takes me to a room to show me a device the team debuted at a major yearly trade event. It looks like any old touch screen, with a picture of a button is displayed on it. When I push it, it feels like something has shifted. But I can tell with my eyes that it hasn't.

"What am I feeling?" I ask. "Is it moving at all?"

"It doesn't matter if the screen is moving," Koehler says cheekily. "What matters is what you feel."

I plead him to tell me and make a few guesses. I ask him if it's just the clicking sound that's giving me the impression of movement, like the old haptics on the Audi. Or maybe it could be a tiny vibration that my brain is interpreting as a click. He refuses to give me any information because it's a button that's still in development. But he has piqued my interest in the new world of engineering touch.

Of all the ways that our products are giving us the tactility that's missing in our lives, the most interesting is our technology because it is precisely what we most often blame for causing our alienation from our senses. So far our devices activate two of our senses: vision and sound. But by incorporating touch, they could bring back physicality to the part of our culture where they could have the most impact. Welcome to the burgeoning field of haptic engineering. Even if the Volkswagen engineers won't help me, I have a good idea of where to turn for answers.

"Think about televisions and computer displays for your eyes, and speakers for your ears," Ed Colgate, a professor of mechanical engineering at Northwestern, tells me at the annual World Haptics Conference being held at his university. "Haptic interfaces are the same thing for your hands. Haptic engineers develop devices that can be programmed to feel and behave like all manner of things when you touch and interact with them."

Colgate is a well-loved, avuncular leader in the field of haptic engineering. He led the first-ever World Haptics Symposium, which became part of the World Haptics Conference, and was the founding editor in chief of the leading journal on the subject, *IEEE Transactions on Haptics*. He explains that the small feelings like rumbling, skin pressure, and muscle strain give us cues about whether we're handling our technology—hammers, cars, you name it—correctly. These signals are missing in many of our virtual interfaces, like when we're using a controller to moderate the speed of a car in a videogame or remotely operating a robot. This places a heavier burden on us to concentrate with our eyes. Engineers working in haptics are working on finding ways to reproduce the most useful of these sensations.

Colgate and I are in a large hall trying out even more tactile novelties. There are phones that let a user send a vibrational pattern, like the beating of a heart, to someone on the other end of the line. There's a dental training simulation that quite convincingly imitates the way it feels to drill into a tooth. Several other presenters are unveiling not fully developed applications but small haptic tricks that they hope to implement one day. One has as stationary block that, when vibrated in a particular pattern, gives the impression it's going from sticky to smooth even though the surface isn't changing at all. Another presenter has me try on a belt that vibrates at my side and makes me reflexively turn at the waist.

The poster boards have titles such as "Active Touch Perception Produced by Airborne Ultrasonic Haptic Hologram," "Should Haptic Texture Vibrations Respond to User Force and Speed?," "The Effect of Manipulator Gripper Stiffness on Teleoperated Task Performance," and "The Effect of Damping on the Perception of Hardness."

The field of haptic engineering originated in the 1940s, Colgate tells me. Ray Goertz, an engineer who designed nuclear reactors at Argonne National Laboratory near Chicago, recognized how dangerous it was for the lab's employees to handle radioactive materials. He built a contraption called a master-slave manipulator that would allow them to stand behind a protective lead wall while they transported the materials with a pair of robotic arms. The master, the part that the employee would handle, and the slave, the attachment that moved the dangerous material, were mechanically connected by push rods. The laboratory wanted to avoid putting more holes in walls, so Goertz had to find another way to provide touch feedback to convey to workers some basic haptic information about the objects they were lifting, including how good a grip they had and how heavy they were. He created a haptic effect by placing motors on the master and slave and gave them an electrical connection. This provided just enough tactile information that the workers could continue to move with precision.[1]

Goertz's model led to similar developments throughout industry. As machines were doing more human jobs that required repetition and strength, operators used an array of haptic effects to control them. Through the years, a number of other nonindustrial uses also popped up at university labs. In the late 1980s a computer scientist named Fred Brooks developed a computer program that could make a robot arm play out the interaction forces and torques of protein molecules as they responded to medications. The readings allowed biochemists to understand previously imperceptible chemical processes to help design more effective drugs. Interfaces were developed that could help people feel haptic illusions such as springiness, bumpiness, hardness, and viscosity and even the feeling of stirring a stick through a bucket of ice. Haptics were used to help people with hand tremors, such as patients with Parkinson's and multiple sclerosis, to operate joysticks by canceling out their bodies' unintentional movements.

Even as haptic experimentation grew, it remained a small and mostly ignored subspeciality of engineering. That is, until a few years ago, when Apple introduced "force touch" to its MacBook laptops and the word "haptics" finally went mainstream. There are sensors in the trackpad that detect how much pressure is being applied, and when they detect the downward force, a motor pushes the trackpad against the finger and vibrates slightly to simulate a click. The finger feels like it is moving down and back up, but it's not. (The Volkswagen button likely uses a similar mechanism.) The pad is also responsive to different kinds of touch. Depending on the number of fingers that are used and how they move, we can change the size of the screen, scroll, and toggle between displays.[2]

This development—using our fingers in multiple ways to interact with a trackpad and with our screens—has opened up possibilities for the field because it required entirely new ways of thinking about how a user could use touch. Engineers are developing a new tactile language that mimics the kinds of maneuvers we make with three-dimensional objects, such as the shuffling of images or the flipping of pages on a flat surface. The challenging part is to make us feel them. Haptic engineers believe these tactile cues matter much more than we realize. They lament the loss the tangible interaction that we used to get when our devices had levers, switches, and gauges, and they're hoping they can bring them back.

Some believe that, as the field deals with these new challenges, it could be on the verge of a golden age. But haptics still remains fairly primitive. The reason is that the distance senses of sight and sound are easier to translate into media. A flat TV screen can show a clear image of a cat, and a speaker can make the sounds of a cat meowing. But the feeling of a cat breathing and purring are almost impossible to replicate on a surface with limited movement, though engineers are trying.

"Haptics is a contact sense," Colgate says. "What matters is the thing that actually touches your skin. And, frankly, it's pretty hard for something to deform your skin the same way that a cat's fur would, unless it's a cat."

One of the conference's most exciting events is a tour of Colgate's lab, which focuses on surface haptics, or creating novel effects using touch feedback on phones and tablets. Along with about two dozen touch experts, I head over the next day. He has a handful of his grad students stationed at tables to show off various iterations of the technology and explain them to visitors. One of the obstacles of a haptics conference is that these presentations can't exactly be appreciated by a large group, since each one must be felt, so long lines form behind each station. We move through at a slow, steady pace.

At the first table a couple students have out what they called Tpads, which are special touch-enabled tablets. Each is opened to a page picturing a variety of black-and-white shapes. At the top is a big square with concentric circles inside it. Underneath is a series of black dots in a honeycomb-like formation. At the very bottom is a dense grid of lines in a crosshatch pattern. When I glide my fingers across them, I can feel them as if they're etched into the glass. This trick is created by altering the speed of the up-and-down vibration of the screen depending on where I come in contact with it. By controlling how much air stands between my skin and the screen, the engineers can alter how much friction I feel, which gives me the impression of contours.

One of the students tells the funny story of how this entire project began. A classmate had brought in a Tide Buzz, a hokey-seeming stain-removing pen that produces ultrasonic vibrations that supposedly help to knock away dirt on clothes. He found that when the vibration effect was turned on, the pen's handle felt slippery, almost wet. As others in the lab started to figure out what was going on between the moving object and the skin and how the brain generates its perceptions around small bits of information, they saw an opportunity. They've since used their knowledge to build up a library of other illusions.

The Tpads can replicate feelings ranging from the bumpiness of snakeskin to the graininess of sandpaper. Once the lab perfected these, they discovered they could even produce three-dimensional effects such as bumps and dimples. People perceive slopes and dips on a surface primarily through a change in the drag on their skin. If friction across the screen goes slowly from high to low, a user can feel like their finger is falling downward into a shallow hole. This feeling is created when a left-to-right vibration is added to the up-and-down motion used to create flatter-seeming textures.

At another table a student has an ultrasonic motor. When voltage is applied to it, it generates a traveling wave pattern across a brass plate that increases friction moving in just one direction so a finger actually gets pushed in the other. This can be used to create all kinds of powerful illusions. It can lead a user along a squiggly line and pull them back if they veer away. It can enhance the illusion of a hole by making the user feel as if they're being sucked down a vortex. The students display a Tpad screen with the image of a light switch that uses this effect. As I toggle it on and off, I can sense it resisting against me before it snaps into place, just like real thing.

The displays have varying levels of persuasiveness. The hole illusion only somewhat works. The light switch is easily the most convincing. But the engineers recognize that they can't strive for perfection. It will never be possible to replicate the feelings of true objects in the world. They can only try to get as close as they can. Through this work, they can learn about touch in ways that aren't possible in biology. Whenever we feel anything out in the real world, many touch sensors are triggered all at once. It's hard to figure out how they contribute separately to our larger tactile awareness. With a screen, it's possible to use one type of stimulation at a time—vibration or friction—to see how they contribute to our impression of shapes and provide information that help guide our actions.

The science is fascinating on its own. But for this research to move forward, the lab members have to prove that this technology could have worthwhile uses. Some futuristic applications for surface haptics are to feel a cashmere sweater before buying it online, to add another layer of realism to virtual reality displays, or to make typing or drawing on a cell phone easier by mimicking the feeling of paint on paper. None of these is fully achievable yet. In the meantime, given the minimal capabilities they have, the grad students in Colgate's lab are trying to show people what the tricks they've created in the lab can already be used for.

Among the questions they're asking themselves are, How can touch change the relationship we have with our machines? How will it affect how we use them? What kind of applications would show off the potential of this technology? This is a serious design challenge. Touch doesn't have a clearly defined artform, such as painting, music, perfumery, or cuisine. It has massage or maybe dancing, although neither is a perfect equivalent. The reason that it's so difficult to come up with an aesthetic form that is solely about touch is that touch is an omnipresent sense that's located throughout the body. It is hard to think about preserving its power while translating it to a limited surface. It's even trickier that we have only a narrow set of haptic tools to work with.

Colgate's research group doesn't have to grapple with this conundrum alone; they created the Tpad Tablet Project as a way to crowd source ideas.[3] They built a website providing instructions for engineers to make their own tablets at home with tools from any hardware store and encouraged them to create their own programs for it. They even held a haptic-athon, in which people came together to build programs using haptics in a single day. They tell me groups have used the Tpad to create guidance systems using virtual nudges for the blind, textural displays to enhance children's books, and apps that take and send photos that incorporate textures.

I think about what I would create, if I were given the chance. And then I realize that I have my shot right here. I ask the students what they think about me taking a couple phones and figuring out for myself what the haptics could be used for.

They happily agree. But I'm going to have to wait a few months because they already have a few loaned out. During those months, I puzzle over the problem.

All of these on-screen haptics are possible because they exploit tactile illusions, says Vincent Hayward, a professor at the Institute of Intelligent Systems and Robotics in Paris, when I meet him in his office. Just like the optical illusions we're more familiar with, these also occur when the brain makes errors. Picture the haptic information going up to the brain like a river, and it flows fast. Although it's full of useful tidbits, there's also plenty that can be ignored. The brain can't go through every single inch of this information river to decide which is worth picking up, so it has to make its best guess using patterns it has seen before. It's a smart strategy for efficiency's sake but also because it's flexible. It's the reason that we can immediately feel two doorknobs that have different designs and recognize they are the same basic object. Inevitably the brain runs into problems when it relies on generalizations. It might gloss over competing information or ignore something important.

A lanky man with the same moptop haircut that The Beatles once wore, Hayward has the eccentric vibe of a mad scientist. His office is filled with even more examples of haptic illusions, which he uses as the basis for his own haptic technology inventions. He lays out a series of three blocks and asks me to rank them by weight. They stand in a row from smallest to largest. I pick up the largest first and am struck by its heft. It looks like it's made of wood, but there's clearly some kind of heavy metal at its core. Next, I hold the medium block, which feels even heavier. The smallest is even more so. With confidence, I tell him my answer. Smugly, he pulls a small scale out of his drawer and places each block on it. They're all about 120 grams.

"They are roughly the same," he says, teasing me through red, round-framed glasses.

"That's weird," I say, picking them up one by one again. "It's so obvious to me that the smallest is the heaviest."

Hayward tells me that this tactile illusion was discovered in the 1800s, and it even works on the blind. Although we don't have a definitive answer for why it happens, he believes that when we see a smaller object, particularly one that is made of the same material as a larger one, we expect it to be lighter. When our assumption turns out to be wrong, the brain majorly overcorrects and perceives it as being heavier. We continue to feel it even though we know we're wrong.

Most of us have seen the optical illusion in which two lines of equal length are placed in the shape of a capital T. Inevitably the vertical line looks longer. Our brains have a tendency to overestimate the length of vertical lines, particularly if we grew up in an urban landscape with lots of tall buildings. Also, bisecting a line can make it seem shorter. Hayward shows me that this exact illusion also works for touch. He makes the same T shape in front of me with wooden dowels, and when I touch them with my eyes closed, the vertical one still feels longer. This is an instance when touch operates the same way as sight.

Next he has me stroke the sides of a pen with my pointer and middle fingers on either side, which feels exactly as I expect it to. Then he has me cross my fingers like I'm telling a lie and do the same thing, so my fingers are on the side opposite to where they were before. It feels like I have two pens in front of me instead of just one. I'm so used to exploring with my hands in one way, so when I change up my approach in this small way my brain gets confused.

Hayward suggests some other tests. I should have another person hold their arm out while they look the other way then tap three times on their inner arm and three times near their wrist. Instead of feeling the taps in those two isolated spots, they might feel like a critter is crawling down their arm at a steady pace. This illusion, called the cutaneous rabbit, has been shown to work even when the center of the arm is anesthetized. The brain fills in information even though the skin can't feel. The reason: we've learned to predict a particular pattern of movement across our arms, so we'll be convinced it's there even when it's not.

There's another popular illusion that involves proprioception as well. Touch the tip of your nose with one of your hands and hold a vibrating massager up to the bicep of its connecting arm. Because the vibration activates the stretch receptors in the muscle, they can be fooled into thinking that the arm is moving, giving the sensation that your nose is growing. This is called the Pinocchio illusion. There's some variability to whether people feel it, though. Just as with visual illusions, how well tactile illusions work is highly variable from person to person.

These illusions are examples of flaws in bottom-up processing to the brain, meaning that a river of information flows up and the brain must decide what it chooses to pick up on. But this makes it seem like the brain is just passively taking in touch cues, and it's much more powerful than that. There are even more illusions when considering another process called top-down processing. This is when the brain determines what kind of information it's going to allow up in the first place. It can actually say "hold the pain," and the nerves listen.[4]

You've probably noticed this happening before. You go to an important job interview wearing a stiff new pair of loafers. They look good, and you spend your day being brilliant in them. But as soon as you leave the building, the pain takes over. You can't take even a few steps in them without cringing in agony. Your feet were likely covered in blisters all day, but it's only now that your body is letting you feel them. Your brain, knowing you had other priorities, decided to keep this piece of information from you. Only when it knows you'll have time for it does it make you deal with the consequences.

Pain isn't the only sensation that the brain can block off. Feeling something passively as it grazes your arm could actually tamp down your response in comparison to actively focusing on it. This is why noticing a puppy's soft coat as it races across your arm is different from nuzzling up against it on purpose. Top-down signaling is also why you can't tickle yourself. Before your fingers ever hit your skin, your brain will dull the tickle sensation. If you're a heterosexual guy, a touch you think comes from a man is shown to feel less intense, less whispery soft, than one that you believe is from a woman. It's worth noting that it doesn't matter if it actually *is* a woman's touch. The same is probably true if you're a heterosexual woman responding to a man's touch.

In short, you only have the comforting fantasy that you're experiencing absolute reality as your body interacts with the world around it. But touch isn't any more truthful than the rest of the senses. That is worth remembering next time you feel an ant crawling up your arm or a tingle running down your spine. What you're actually experiencing through touch is just your brain's subjective version of what it thinks you need to know, based on years upon years of learned biases. The miscalculations of our tactile systems are rich material for engineers.

"You can create feelings of sensation where there's nothing there," Hayward says.

Hayward is curious about how the brain's flaws can be used to enhance the way we use our technology, aside from sheer novelty. So far, there are a few clues. There is some limited research that shows that haptics can help intensify our emotions when we're playing videogames and make us more engaged when reading electronic books.[5] The addition of a sensory component can click on parts of the brain that are usually not involved to make such mental exercises more immersive. This is helpful when using flat screens. But the best way to conceive of what they add probably isn't new technology, where it hasn't been fully explored. It's more helpful to think about what's now missing from the objects of the past.

While our technology is presented to us as ever improving, a generation that grew up with digital tools is now seeing renewed value in analog. Take books, one of the most-studied examples of old versus new technology. Many studies show that how we absorb material from e-readers is different from cloth and paper books, and a lot of it has to do with the physicality of the page. We think of reading a book as an abstract cerebral activity, but our brains actually process it similarly to traversing a physical landscape. There's a left page and a right page, we turn each one by one. We're able to monitor our progress as we go by feeling the stack of paper on one side getting smaller while the other gets bigger.[6]

The tactile details are important. People report having a firmer grasp on what they read on paper, in part because it sticks in their motor memory. They're better at recalling, for example, general locations of key phrases or scenes. Students who are posed questions after reading a passage can flip through paper faster to find key sections. The sense of forward motion isn't as obvious in digital reading material, which display only a progress bar and percentage number at the bottom of the screen. We can't navigate them as well, which is part of the reason using them is more taxing.[7]

People say they simply enjoy the experience of engaging with physical books more. There's a certain style and weight we expect when we're thumbing through *He's Just Not That Into You* and *Ulysses*, and our brains enjoy it when what we feel fails to match these predictions. A Kindle makes all books feel uniform. The permanence of ink on page, versus the ephemerality of the images of an e-book, might also make some of us subconsciously take the reading experience more seriously in print. We are more likely to make the effort to go back and reread complex passages. When we read actual books, it's also easier to put our stamp on them, dog-earing pages or writing notes in the margins. We get a sense that we're having a unique and personal experience with the material.[8]

Writing is another practice that's markedly different in analog versus digital. There's something meditative about the act of putting pen to paper. It's a sensual experience—the pinch of our fingers gripping the pen, the metal gliding across the smooth page. Writing forces us to slow down, our minds tethered to the pace of our bodies. Just as with reading, it is like moving along a rolling, looping, rugged road. We don't have to think as we draw out the letters, as we've been doing since we were very young. It's this act of repetition that puts us in the mental state we need for the words to spill out, one after the other, providing rhythm and flow needed to elicit insight.

Studies show that children who have not yet learned to read or write are better able to identify a new letter correctly on a computer if they've first written it out in freehand, versus seeing it on a computer or tracing a dotted outline.[9] Older students retain information better when they record lecture notes by hand instead of typing them. Psychologists often talk about the difference between knowing and doing, which explains this difference. The latter is more difficult because it requires us to carry out our ideas and exert physical effort. But it's also likely to stick with us longer.

The reason we continue to prefer some of our older artifacts isn't just about nostalgia or street cred. We're gravitating toward touch. We may not be aware of all the tickles and bumps we get from physically handling objects, but in subtle yet significant ways they affect how we think. They activate parts of our minds associated with action and get us to engage with them in a different and deeper way. Adding more tactile effects to our phones and computers could do the same. They can make us feel more connected to our devices and trigger us to manipulate and play around with them in new ways, perhaps even expanding the possibilities of what we can create with them.

This is what's on my mind over the months following the World Haptics Conference. But it's not what I'm thinking about when I finally do get the Tpads mailed to me at my new home in Fredericksburg, Virginia. In the past few weeks, my life has been a whirlwind. I've moved here for a job teaching journalism at a university south of Washington, D.C. Getting the job offer set off another course of events in my romantic life. My boyfriend, Kartik, had supported my taking the position, but as my move came closer it was clear he was having doubts about joining me. He was worried about what it would mean for his own career, but we didn't talk about it that much. Instead, we got into many passive-aggressive fights. After one of these, we broke up. Then, a couple days before I left, we decided to change course and get engaged instead.

As I'm sitting alone in my stark new faculty apartment, surrounded by the standard-issue dorm furniture, I obsess over everything that's happened between us. I wonder if we made our decision too fast and whether Kartik felt forced. Maybe it was all a huge mistake. But I don't tell him any of this. Instead, we talk about how much I'm enjoying getting to know my colleagues and students and about the Pinterest pages I've been checking out for inspiration for our wedding. We've never been good at heartfelt conversations over the phone. With the Tpads in hand, I'm excited to have a project to work on with him, something to get our minds off of everything that's happened between us. And I'm not just interested in solving engineering questions anymore; I want to know how communicating through touch could help bridge the distance between us.

On my next weekend trip up to New York to visit him, I give him one of the phones and tell him that we'll be performing an experiment together. By sending haptic messages to each other for a few days, he can help me figure out what touch might contribute to our technology. There's a program on the phones that lets us send haptic pictures. It turns the black-and-white images we take into a textural pattern based on their lightness or darkness. The first week I get some images of the wood grains on his desk and the creases of his hand. They're kind of interesting to feel. His favorite of mine is a picture of my pajamas, which are covered in black-and-white snowflakes. The stark pattern creates a texture on the screen that is almost like softness when he touches it.

"Do you think touching them helped you feel closer to me?" I ask.

"Well, the pajamas reminded me of you," Kartik says. "I don't really think it was the haptics."

I don't think the images are very compelling either. The experiment fizzles out when we stop taking our phones out with us because they're bulky and we're hardly sending each other anything anyway. I'm disappointed, and I end up talking over the issue with my new colleague, a computer science professor. Together we come up with a new plan. What would be even better than the static haptic pictures would be the ability to touch Kartik virtually in real time. My colleague says he can have his son, a computer scientist based in Austin whose hobby is creating prototypes for smartphone apps, make an app for us. He has it ready a week later, and we agree to test it out together at first.

When I swipe over the image of a finger on the smartphone display, it feels as if he's touching me back. His finger doesn't have the texture of a real one, but it does seem ever so slightly raised, like he's reaching out at me. I keep running over it, amused, because I can see with my own eyes that the glass is completely flat. The vibration of the screen's surface is changing imperceptibly depending on where my skin makes contact with it. My brain interprets this tiny shift in friction as a three-dimensional bump. It takes me a few seconds, as I move my finger back and forth, to remember that I'm having this exchange with an actual human, and I stop abruptly. I wonder if he's as creeped out as I am.

I'm as uncomfortable as I would be if I were fondling a real stranger in the flesh. I don't know him well enough to ask if he feels the same way, but the stiltedness of his speech and the speed with which the image of his finger disappears when I'm done suggest that what he's experiencing is not too dissimilar from me. I thank him profusely and get off the phone quickly. I actually take our mutual aversion as a good sign. If this short interaction could elicit the same kinds of emotions as true skin-to-skin contact with someone we don't know, then I'm hopeful that the same might be true when I use it with Kartik.

Kartik gets the phone in the mail soon after, and we test it together. Our fingers meet up near the center of the screen, and I can feel the little raised spot, like the glass screen is a little warped. We stay there, rubbing the impression of each other's presence. I let him take a few seconds to make sense of what he's feeling and then I move my finger away. He tries to come closer again. I dart away, and he chases me. I stop and let him catch me, letting his finger linger on top of mine. We gently nuzzle them against each other. We are flirting, haptically!

Some of the greatest love stories of all time started with the gentlest of touches. When touch is forbidden, it takes on increased significance. Anne and Wentworth in Jane Austen's *Persuasion* were once a couple but broke off their engagement because of opposition from her family. When they enter each other's lives again, it's through nuances, including in the way he touches her, that she can sense the feelings that are still there. "It was a remainder of former sentiment; it was an impulse of pure, though unacknowledged friendship; it was a proof of his own warm and amiable heart, which she could not contemplate without emotions so compounded of pleasure and pain, that she knew not which prevailed."[10]

In Patricia Highsmith's *The Price of Salt*, later adapted into a film titled *Carol*, a young woman named Therese, who works at a department store, is trying to explore a relationship with an older woman at a time when same-sex relationships happened mostly in secret. An accidental touch between them feels almost electric. "Once the backs of their hands brushed on the table, and Therese's skin there felt separately alive now, and rather burning."[11] In the dystopian universe of *The Handmaid's Tale*, Offred softens her skin with butter because she isn't allowed lotion and thinks about what happens to the human spirit when deprived of touch as she starts a relationship with a driver named Nick. In *The Black Prince*, Iris Murdoch writes: "Only take someone's hand in a certain way, even look into their eyes in a certain way, and the world is changed forever."[12]

Okay, so maybe it's an exaggeration to say this haptic touch between me and Kartik is anything like those examples. It doesn't feel anything like the intimacy of skin-to-skin contact, but it is surprisingly moving. For the first time in a very long time, we're allowing ourselves to connect. Instead of being burdened by hashing out what's happened over the past few months, we're enjoying each other's company the way we did when we had just started dating. We realize we need more moments like this one. When our emotions are at their most intense and the words haven't been coming, it is through touch that we are able to express ourselves again.

Soon after our experiment, we have to give our phones back to the lab, and I miss them immediately. It's akin to someone telling me that I can keep my smartphone, but it won't have a camera or a voice recorder anymore. Having that extra ability opened me up to new ideas about what my technology can do for me. What I had once thought of as a needless accessory had become an essential communication tool. That's what so often happens with our gadgets. We don't realize we need something until we have to give it up.

In his 1932 novel *Brave New World*, Aldous Huxley writes about "feelies," movies that stimulate more than just the eyes. One character in the book recounts with awe how he could feel the bearskin rug under him as he watched the main characters having sex on the screen. In the book, these films are used as a propaganda tool, a way to keep people brainwashed in a sexual, materialistic culture. After John the Savage, a character born naturally of a mother and a symbol of the old world order, watches his first feely, he is unimpressed by the gimmicks, which he believes mask the film's lack of emotional complexity.

Huxley predicted this haptic future, but he viewed it as a sign of materialism and hedonism. He passed it off as a cheap technological trick, just the latest commodity for novelty-seeking consumers. But Huxley is wrong, which isn't to diminish him. He was just presenting an oft-repeated belief about touch, that it's animalistic and lustful, a perspective mired in all kinds of cultural assumptions about our lowest sense. What he didn't present was the other possibility that we're now seeing in our emerging technology, that touch provides important feedback that makes our interactions with it more intuitive and could give us the kind of connection that we're all craving.

Because haptics is now introducing touch to products that have been without it before, we can see with fresh eyes how touch contributes to our lives. Small haptic effects have long eased our interactions with our machines to make them seem like extensions of ourselves. For years we have used them to move beyond the limitations of our bodies, to make repetitive clean motions and lift heavy or dangerous objects. When we add the same kinds of cues to our interactions with our phones and computers, we see a similar effect. Very basic haptics can help to make the way we manipulate our screens more precise. For example, they make it easier to hit the right letters when we're typing or to toggle between various pages.[13]

But this is just replacing the functionality of the moving parts of the past. We don't yet know how haptics will inspire new ways for us to interact with our screens. We may use these effects in even more profound ways than we have been able to before, not just to perform tasks that require more strength or agility or are dangerous but to enhance our emotions and feelings. The same attachment we have to books and writing on a notepad or listening to vinyl could get translated to what we feel when on a phone or computer. They'd feel more like solid objects than an endless spectacle of images.

We talk a lot about how technology distracts us. Whenever we're engaged in a demanding task, such as learning a new language, reading a book, or doing our taxes, we have to stay completely focused. But that's harder when we're yanked away by a blinking phone that has its own agenda. Our technology is designed to make us abandon our plans and follow it. While we've got plenty of applications to keep us from checking email or to reward us for completing specific segments of our work, we don't usually think about how we could use the senses differently. We know that navigating physical objects can make us more absorbed when we're doing difficult work. Adding more manual functions to our phones and tablets could make them hold our attention on our work for longer.

In the same way, the physical engagement that this technology offers could make us less distracted when communicating at a distance. Part of what made my experiment with Kartik so important to me was that it was the only time I was fully present to him when talking on the phone and not folding laundry or sorting through my papers. We both had to be present because we were working on something together, which forced us to reengage. When we're quickly shooting off text messages and posting on social media while going about our days, we might have the impression that we're communicating, but we're missing out on the possibility of true intimacy. It could be something as simple as adding a tiny haptic vibration that can make these interactions feel more robust.

In the time following my experiment with Kartik, the research on surface haptics continues to inch along slowly. Small improvements are made each year that add another touch of realism. The truth is that, just as with predictions about virtual reality, the possibility of re-creating the sensations of real life are likely overblown. In the case of touch, there are even more challenges because of the anatomical actualities of our bodies. However, it's still useful to think about how even that limited addition of touch can get us to engage and create differently with our technology. And by making us focus on touch in a new way, it reminds us to appreciate the vibrancy of the sensations that float by us every day.

9

Tactful

Building Machines That Can Touch Us Back

Gerald Loeb had never studied touch when he was asked to join a Department of Defense program called Revolutionizing Prosthetics. Its purpose was to produce high performance units for patients with amputations from the wars in Afghanistan and Iraq. As a biomedical engineer at the University of Southern California, he was responsible for just one part, which was developing algorithms to represent the movement of the musculoskeletal system. But at a group meeting, as he listened to the experts assigned to other issues—the comfort of the socket, the battery, the motor, the control system, and particularly how to add feeling—he started to think that the way they were going about the task of replacing sensation was all wrong.

Several participants kept alluding to the idea that the skin acts as a thin sheet covered with little points of sensitivity that can map out the contours of whatever it comes in contact with. Even as Loeb was sitting there, he shook his head. He considered his own body, and it seemed obvious to him that it was the pattern of stretching and tightening throughout the flesh that matters, not just the sensors on the surface. There needed to be some kind of communication system from sensor to sensor for the prosthesis to have a full picture of what it was touching. Not to mention, if sensors were needed all over the prosthesis, they could easily fall off or get damaged and the unit would need constant, expensive repairs. Later that evening, at a bar with a friend, he examined how fingers really work, and on some napkins, he sketched out a few initial ideas.

This was the beginning of a two-and-a-half-year obsession for him. By the end of it, Loeb had created the mechanical finger he envisioned that day. It consists of an inner core resembling a bone that is coated in electrodes. Surrounding them is salt water encased in a smooth skin to prevent evaporation. Just like in our own fingers, when the skin grazes across a surface, the entire structure feels it. The electrodes measure that pattern of deformation and provide a reading of what they sense. Loeb started a company, called SynTouch, near his college campus to improve upon the model and manufacture these fingers. Soon these so-called BioTacs, which are some of the best replicas we have of the sensitivity of our own appendages, were in use by companies and labs around the world.

At an office in a loft above a grungy strip mall, Loeb and his former student Jeremy Fishel, who was hired as chief technology officer, go over the enhancements they have made to the original model so it has more capabilities than just detecting pressure and, more and more, works just like an actual finger. They installed a small gauge inside the core so it could measure vibration. They added a thermistor at the tip of the finger to measure when heat flowed out. As they made changes, the prototype started to look more and more anatomically correct. It even had fingerprints, which made it more sensitive to vibration, and a fingernail, which helped to locate the pattern of the finger's movement better.

"Biology is smart, so we just copied everything about biology and threw it in there," Fishel says. "It's a very complex structure in a big squishy skin. The fact that the skin is nailed down with a fingernail matters. The fact that the bone underneath has a rather complex shape matters."

They show me the latest version of the BioTac with its top layer of skin removed. It feels a little like a water balloon filled to its limit, and through it I can see all the parts inside—the bone-like rod at the center and the temperature and pressure sensors. Fishel opens a program on his computer that displays all the measurements that the finger can make. He swipes the BioTac across the trackpad of his MacBook Pro, which is made of glass, then on the brushed aluminum of the rest of its body, then onto the wood desk nearby. The data visualization shows how dramatically the texture and the temperature change on every surface. The fingers are even more sensitive than our own.

Of course, re-creating our sense of touch involves more than just an imitation body part. To function fully, movement is needed. SynTouch engineers have gone through the process of observing how people explore objects and have thought critically about the qualities they look for. How do their fingers have to move? How much force needs to be used? How fast do they slide over a surface? Do they accelerate or slow down? They have translated those findings into an efficient set for a robot to perform. Through making these movements, a robot can tell pretty accurately the difference between one texture and another.

SynTouch's BioTacs are being used by other research groups for functions that are even more advanced: to catch balls, stack plastic cups, type on a computer, spoon food from a bowl, and play Jenga. Having both tactile and visual feedback helps maneuver fragile objects more accurately and gives them more dexterity. But so far, such machines still have to be told how to use their newly sensitive fingers. Eventually, SynTouch foresees that a computer will not need to be programmed to perform each of these actions separately but will actually be able to process touch cues and explore on their own. If they're given a complex tasks, they will be able to see and feel through how to accomplish them. This will require that their newly sensitive robot bodies be hooked up to a smart robot brain.

So far, we've gone over just one aspect of haptics, which is focused on creating tactile tricks for our fingers. This other branch is trying to do something even more ambitious, which is to make machines that can feel for us. If what I've seen so far is touch 2.0, this is touch 3.0.

We don't yet have artificial intelligence sophisticated enough for computers to act autonomously. But in one lab, touch cues interpreted by the BioTac are already being transmitted straight to human brains in experimental touch-enabled prosthetic limbs. It's the best example we have of what this technology will look like.

On a damp fall day I visit the lab of Dustin Tyler at the Louis Stokes Cleveland VA Medical Center. One of his test subjects, an amputee named Igor Spetic, is already there, sitting amid a flurry of grad students. The sleeve of his bright blue polo shirt is bunched up to reveal wires sticking out of his upper arm, which are being connected to a prosthesis. When the system is all in place, if the BioTac fingers of Spetic's prosthetic hand sense the texture of what they're touching, a corresponding electrical wave pattern will get sent to the wires, which are attached to electrodes that are surgically implanted around the nerves of his amputated arm. This means he registers the sensation as if it is a very rudimentary form of touch coming from his missing hand.

In a scene that makes me think of Frankenstein's monster, Spetic leaves his arm outstretched for them to fiddle around with as he matter-of-factly tells me the details of the accident that caused him to have his hand amputated. He says it was just a usual day at the factory where he worked as a forging hammer operator, which meant that he used a machine to pound sheets of metal into parts that could be used for purposes such as hip replacements and military equipment—until something went wrong. Either his clothing caught or the machine went haywire, and his hand was crushed. He never bothered with an investigation because it wouldn't bring back his hand, and he didn't want to get stuck in the past.

He prefers to think about the future and how being involved in this research could help others. Spetic knows innately the importance of the sensitivity of our hands. Before the accident, he worked as a machine operator for decades. He prided himself on the physical nature of the work he did at the factory, which required a type of skill he valued over the book learning he was required to do in college before he dropped out because tuition was too high. His self-worth revolved around wrangling finicky machines. He has learned to express himself primarily through his body, the way many of us are more comfortable doing through language. He wants these abilities back, which is the reason he keeps coming back to the lab week after week.

The project is still at an early stage. There's only a limited set of sensations the experimental prosthesis can administer so far, which the lab workers have discovered through trial and error by applying various patterns of electrical stimulation to a few contact points on Spetic's nerves. They can create pressure and the sensation of cotton balls, water trickling, and sandpaper but aren't able to re-create the full range of natural-feeling touch yet. For that, the lab is collaborating with other specialists in the neuroscience of touch to build up a library of sensations for Spetic. In the meantime, they want to understand in a broad sense how adding some sensation can allow him to use his prosthetic arm differently. This day in the lab is preparation for the group's most important experiment yet, during which they'll allow Spetic to take the prosthesis home for the very first time.

Once he's all set up the students begin performing the tests on him. They blindfold him and make him wear headphones playing white noise, so he's operating based on what he feels alone, then have him pick up a series of foam blocks of three different densities, from spongy to stiff, with the stimulation of the prosthesis turned on and off. When it's on, he can sense the pressure of the blocks. Next he has to take clothespins off a rack and line them up on a metal rod while naming any fruits and vegetables he can think of, again with stimulation on and off. He has a quiet determination as he goes about the tasks. As he rattles off his list—strawberries, raspberries, cherries, tomatoes—I try to do the same in my head and helplessly stumble. By the time he is done the lab workers are equally impressed.

"You're a rock star," they cheer on.

But the results of both trials show that there might be no benefit to having tactile stimulation. Even more surprisingly, no one in the room bats an eye. They say that in the past they might have been bothered. But over time, they've seen that the benefits of touch aren't always apparent in a lab experiment. The engineers have deduced over time that there are two main areas that the touch sensations of the prosthesis help with—in handling delicate objects and in multitasking so he doesn't have to be hyper-focused on one goal at a time. One of the most successful lab tests, which became popular on YouTube, shows Spetic blindfolded and wearing headphones while he pulls cherries off their stems. Without touch, he repeatedly squishes the fruit. With it, he gets the pressure right every time.[1]

Another unexpected benefit of the touch-enabled prosthesis is that Spetic's phantom limb pain mysteriously went away. The pain used to be excruciating, but it started easing up during the very first day of testing. The reason is that the stimulation provided his brain with its first glimpse of information from his lost hand. Before that, in Spetic's brain, his hand was still in the same position it had been when he lost it—clenched in a fist being crushed by the jaws of a machine. His brain couldn't let go of that memory until it received sensation from it again. While there are other, simpler ways of relieving such pain, like with a mirror, the finding does show that the prosthesis does make him feel like the hand belongs to him in a way other prosthetics without sensation don't.

But what Spetic says he enjoys most about having even limited touch cannot be quantified in the lab. He just likes the impression that he has his hand back; it's not about functionality but about embodiment. In between tests the engineers have spotted him moving the hand around the table just to amuse himself. He often takes longer with tasks when the stimulation is on because he's paying more attention to what he's sensing. Even if he is slower, he is more confident with each move because of the feedback he gets. Using the prosthesis doesn't give Spetic anything like the true touch that you and I feel. But even whatever rudimentary stimulation he is given is meaningful to him.

"Touch is about connection, the way we explore the world," Tyler says. "It's a big part of making us feel like part of our surroundings. As engineers, we tend to underestimate the psychological aspect, but we shouldn't ignore it."

When Tyler speaks in these kinds of general terms, other scientists tend to react skeptically. If this prosthesis ever becomes widely available, it is likely to cost tens of thousands of dollars because it would have to be customized for each user. For insurance companies to cover any part of these expenses, he has to show them that it can help people return to work, either sooner or with more skills. There are easier ways to go about that than surgically implanting machinery in patients. In one SynTouch lab experiment, for instance, information from the BioTac was transmitted to the forearm of an amputee rather than the nerves. This provided enough sensation to help with some accuracy of movement even though it didn't restore touch. What such an option didn't accomplish, however, was giving its user an attachment to the prosthesis. "There's a benefit associated with having that sensation that's harder to quantify and convince people about," Tyler says.

At the end of the day in the lab, the entire team walks Spetic to his car. Tyler has in hand a new DSLR camera; like a proud father sending his kid off to prom, he asks Spetic to pose along the hallways of the center and in the parking lot. Everyone looks exhausted from late nights trying to ready the device for this major test, although excited that the home trial will give them a much clearer view of how the technology could help him. Spetic, even after his mental workout, looks energetic. He's excited that he can take his arm home with him. When he has had to leave it behind in the past, he's broken down sobbing at times. And he has one plan in mind for how to use it.

"I'd love to hold my wife's hand and actually feel it, not just feel the weight of her hand holding my prosthetic," Spetic says. "It would be nice to feel fingers. I know I won't feel her skin or anything like that. That human connection would be great. I'll probably be a blubbering idiot."

I call him up after his home trial, and he says that he definitely noticed some differences when he was using a unit with sensation. His usual prosthesis feels like a tool to him, like one of those metal grabbers that park workers use to pick up trash. The experimental one felt like it was a part of him. Without any conscious effort, when he was grocery shopping, he reached out with it to pull cans off the shelves. When he was chopping vegetables at home, the same thing happened. He wasn't worried about squishing their soft flesh with his fingers, something that he'd normally have to be extremely careful about. "I didn't even notice I was doing it at first," he says. "It was just easier to do without thinking about it."

Unfortunately, there were a few glitches along the way, including one that made him have to turn off sensation completely for almost a week, and he didn't get to touch his wife's hand, but he's happy to save that for next trial run. He did wear the experimental hand to his classes at Lakeland Community College, where he studies mechanical engineering. His fellow students and his professors hovered around him as he showed them how it worked. They took turns shaking hands with him. With his usual prosthesis, Spetic would have been too nervous to latch on too tight and hurt someone. He usually just holds out his open hand for someone else to hold onto. But with the touch-enabled unit, it was totally different. He squeezed, and the sensation let him know exactly when to stop, which let him feel more engaged and present. As the research has progressed since his home trial, Spetic has finally gotten to touch his wife's hand, and it was just as meaningful to him as he had imaged it would be. "My confidence was way more," he says. "Way, way more."

Spetic's story teaches us that it's because of our sense of touch, particularly in our hands, that we can use our bodies to spontaneously express our will. Some anthropologists say that this hand–brain connection is one of our most human of traits. In his second book, *The Descent of Man*, Charles Darwin examined the anatomy and behavior of humans to endorse a close evolutionary relationship with African apes. His main question was what set humans off on their path to differentiating themselves from their common ancestor with these primates. His answer, to put it simplistically, was bipedalism. Darwin posited that "owing to a change in its manner of procuring subsistence, or to a change in the conditions of its native country," a member of the primate order came down on land and walked on two legs.[2]

By this point, our ancestors already had big brains. They were using very simple tools, for instance, sticks to dig into tree trunks for insects or stones to throw at their enemies. Two-legged life freed up their hands so they could use them more wisely. They made more advanced tools, which were necessary for their more resource-scarce and predator-rich landscape. These tools then gave them access to new types of foods like meat or seafood and to cooking, which provided more nutrition. In Darwin's version of the story, our brains gave us the advantage of advanced use of our hands, which is what allowed us to thrive as a species.

Darwin's theory at that point was based on conjecture since there wasn't yet any physical evidence, such as bones, available to confirm this idea that bipedalism led to wholesale changes in humans. Years after he came up with it, a fossil nicknamed the Taung Child from 2.4 million years ago was discovered in 1924, and it suggested that a big brain came many years after humans became bipedal. A new timeline of human evolution began to emerge in which the hands were even more central. It wasn't that the brain allowed people to make tools but rather the inverse; the skills needed to live in a dangerous environment, including the use of hands for tool making, helped to propel forward human intelligence.[3]

This idea was supported by even more fossil evidence. The oldest stone tools ever found are about 2.6 million years old, and hominin brains began to expand dramatically starting about 2 million years ago. This means it's possible that it was by tinkering and exerting their ideas on raw materials they found in nature that our ancestors developed the basis of intelligence. This is, of course, just one of many changes that could have led us to become the species we are today. There are other factors, including diet, that contributed to our mental development, but the legacy of the hands and brain working together is still with us today. We can even see it if we look closely.

Brain imaging studies show that when we use our hands, the density of touch receptors in the skin and the sensorimotor area representing them in the brain grow. According to Daniel Goldreich, an associate professor of psychology at McMaster University, you can do your own experiment at home by taking a tool like a pincer and putting the points inches apart. Touch the points to your palm and feel the two spots that are being touched. Then move the points closer together. Notice when the two points start to feel like just one. Practice this task again and again. Even fifteen minutes spent detecting the two individual points can actually improve your abilities, which suggests that the parts of the brain responsible for touch processing are extremely malleable.[4]

It's no accident that people who use a certain body part repeatedly, through reading Braille or playing a violin professionally, have larger mental representations for those areas in order to perceive nuances through their hands. Their physical behaviors actually generate changes in their brain structure. This plasticity could have been the mechanism through which our use of the hands led to broader changes to our intellect. There is a theory that the Broca's area of the brain, which is responsible for preparing for the movements of the tongue during speech as well as the fine motor actions of the hand, reveals the close mental link between tool-making and the development of speech.

The Broca's area is rich in mirror neurons, nerve cells that get activated when we're observing others and learning to imitate them. It began developing around 2.5 million years ago, a time when people were already using stone tools but before there was a rapid growth in the size of the brain, and it could have helped ancient humans notice the kinds of gestures it took to make tools and make replicas. Similarly, it likely plays a role in how we interpret each other's utterances and repeat them through the process of learning how to speak. The Broca's area is extremely active in the brains of babies when they're first learning to babble.

But that's not the only way that our hands could be related to our ability to use language. Robin Dunbar, a British zoologist and author, has proposed that the grooming done by primates could be a predecessor to language.[5] When visiting any zoo, we can see monkeys engaging in this activity, which involves combing through each other's hair, parting it down to the skin and using the mouth to remove parasites. It's an effective way to trigger the release of endorphins, opiates that make them feel lightheaded and woozy and comfortable with each other. It's a very focused behavior, and they spend about 20 percent of their waking lives doing it, which tells us how important it is. It's a method of building coalitions, which is crucial for survival.

As primates evolved over time, from the ancestors of the new- and old-world monkeys to the orangutans and gorillas, Dunbar says they developed more active social lives. The great apes, which include chimpanzees and bonobos, our closest primate relatives, have the largest social groups of all at about fifty members, and they spend hours each day grooming each other. At some point, there's just not enough time in the day to groom anyone else, which is why group sizes had to be limited. To form larger communities, they needed a more efficient way to show their closeness to each other. This is where language would have been useful. It was a way for many people to bond at once. Spontaneous laughter could have led to the development of song and then to a codified system of words and grammar.

It's telling that even though we now have language, we still use touch for many of the same purposes primates do: to build relationships, to overcome interpersonal obstacles, and to maintain group unity. When we're at our most vulnerable and words start to lose their power, touch becomes a vital form of communication. We remember that our emotions are physical, and our hands are our most potent way of expressing them. Dunbar speaks to the limits of language to express raw emotion in his book *Grooming, Gossip and the Evolution of Language*. "It is a most wondrous invention for conveying bald information, but fails most of us totally when we want to express the deepest reaches of our innermost souls," he says. "At this crucial point in our lives, grooming—of all the things we inherit from our primate ancestry—resurfaces as the way we reinforce our bonds."[6]

Scientific theory aside, the legacy of the hand–brain connection is most deeply felt by people who work with their hands most often, such as artists, athletes, and tradespeople. In the book *The Hand: How Its Use Shapes the Brain, Language, and Human Culture*, for instance, Frank R. Wilson speaks to several professionals who say their sensorimotor orientation shapes their personal identities. They say their handiwork teaches them to look outside of themselves and learn about the physical qualities of objects and materials, whether a slab of clay, a bow and arrow, a car engine, or a cello. In order to achieve what they want, they have to navigate the limitations of their materials as well as their own skills, which teaches them empathy and compromise. The sense of satisfaction they get from this physical labor gives them an enormous pride that can't be replicated through purely mental work. They lament it as something that's missing in the modern world and in many people's jobs.[7]

One example is Patrick O'Brien, a guitar player who experienced an injury and had to relearn his instrument and went on to helping others in the same boat. "There is a particular emotional element involved in this kind of work, a way of perceiving things and a need to predict what's going to happen next. It's a very little world—in your hands. Whatever you can do with your hands gives you a small world that you can actually cope with, as opposed to the big world, where perhaps you can't."[8]

Given this background, it's easier to understand why a touch-enabled prosthesis means so much to someone like Spetic, for whom working with his hands has always been so important. When we think about what touch technology can do for such patients, we go about it too logically. But it's not just about enabling them to move objects. If we look back at the evolutionary story of our hands, we see that their sensitivity is a building block of our identity. When there's a feedback loop between the brain and body and we can act on what we want, we feel complete.

What does this mean for the rest of us? The same touch technology used in prostheses could eventually become more mainstream. It could be used to transmit sensations remotely so we could work as deep-sea welders or use tools in war settings without the risk. We could use it to extend the limits of our bodies, which would mean that our robots would no longer be just tools, but part of us. Furthermore, haptic technology that makes it easier to operate our machines is considered just a stopgap on the way to total automation—as in touch-sensitive robots that wouldn't need us to operate them at all. Robots would be able to perform not just uniform movements but also complete complex tasks that require adaptability to delicate conditions and materials.

They would accumulate the skills needed to sew clothes or turn tiny screws or make arts and crafts that require a great deal of delicacy. Driverless cars could handle bumps in the road better than some cab drivers. They could even take over more intimate tasks, like surgery or caregiving. It cannot be overstated what a huge development it would be to give robots that level of sensitivity. Machine learning is, of course, already improving quickly. Robots can win chess games, identify our faces, and engage in meaningful—even romantic—conversations that can fool real people. But you know where they fall short? They can't sense touch. They can't reach into a pocket for a set of keys, find them amid the clutter, and hit the precise button that unlocks the doors—all without looking. This is a sophisticated form of intelligence that we fail to appreciate.

If we can re-create touch in our machines, then we have to think about the human impact. For this it's helpful to take lessons from what we've learned about the value of touch in prosthetics because here, too, we think too practically about what these autonomous robots will be able to do—mostly, which of our jobs they'll be going after first. Of course, we do have to think about the future of human work when more robots can do our existing jobs, whether we'll protect people through job training or universal income. But we'd be wrong-headed if we limited our concerns to finding a new source of income. We need to ask ourselves if the *way* we relate to these robots will change. And what will that do to how we use our own sense of touch?

I picture a time twenty years from now when, with a few more gray hairs, I'll head over to a new, themed café in town, have a machine hand me my latte, and sit down to drink it while playing a game of Jenga with a robot. We'll take turns checking out the structure of tower of blocks to see which one can easily be wriggled loose. With the specialized knowledge of physics developed over months playing at a factory, there's a pretty good chance it would beat me. Its hands and fingers would move with a precision that mine are incapable of. I wonder if I'd feel the need to tell it "good job" or get weirdly competitive or have more fun with it than with a person.

Responsive robots would be safer to share a space with, so we could start interacting with them the way we do with people. We'd be forced to give them our trust and recognize their sensitivity, which means they'd maybe feel like our partners and collaborators rather than our dumb slaves. After all, if it is our senses that give us our feeling of embodiment, then if we imitated the same in our machines, they could have something like a sentient system. Or perhaps because we know they're made up of metal and wires, we wouldn't need to keep up the niceties we do with human beings. We could treat them without much regard, and they wouldn't expect much more of us, which would mean we'd get less practice in basic kindness and empathy than we usually do.

You see, how we think about our robots matters to who we become. We will have to consider what we prefer—an increasingly frictionless existence or one in which human feeling matters. We would also have to look at ourselves and ask: what are robots good at, and what are we good at? Is it vision or touch? We could continue to place more value on our intellectual abilities, and we could outsource touch to our robots or we could do the opposite, which would reverse our understanding of the senses for the first time in recorded history. We could find more fulfillment in making art and cooking and caring for the sick and in hospitality, the kinds of activities that have been so critical to human development and connection in the past. We might decide we would prefer to meet other people in person again. Our ability to feel would be more important than our intellect. This will be a crucial crossroads.

The arguments for embracing touch come mostly from the rhetoric around screen use, since much of it focuses on how overstimulation of our eyes rather than the studied use of our hands is affecting our brains. We talk about how screens give us trouble sleeping, get us hooked on hits of dopamine that affect our ability to focus for long periods, and make us gain weight. In sum, our bodies become dysregulated, and we're driven by immediate concerns over our natural rhythms and instincts. These effects are uncomfortable for adults, but they're particularly impactful to children whose brains are just developing. Using touch is simply healthier for us.

Psychologists say that children who learn virtually how to throw balls or paint and color aren't able to do the same in life. When they interact and play in physical spaces, cutting up cardboard boxes, working out jigsaw puzzles, throwing a ball, and fighting with other kids over toys, they receive a tangible reminder of their own existence and the fact that their surroundings won't always conform to them. They come to understand where their boundaries lie and how to navigate in creative ways to achieve what they want. Failing to learn these important lessons, by playing videogames instead, they can be less patient and have a harder time adapting or compromising.

Some educators advocate for more hands-on learning in the classroom. It's a much bigger challenge to grasp how to play a few chords on a musical instrument or to take depth into account in a drawing than it is to memorize material. But it's more rewarding because students can see real results, not just a score on an exam but as a new ability to take out into the world, and the lessons stick with them. After a while the act of repetition produces the kind of effortless movement that helps to lead to a mind state we call "flow." We're very focused on giving kids practical knowledge through a text-based, eye-oriented approach, but what we're missing is the need to pass on an appreciation for attaining physical skills, which is known to lead to lifelong learning.

For all of human history, our brains have patterned themselves based on the environmental inputs we give them. In tracing back our evolutionary story, we could realize that the sensitivity of our hands is so important to us that we're going to find ways to engage with it again. Fiddling around and making things has always been our gateway for solving the larger intellectual issues we face. On the other hand, we could continue on our eye-focused path and end up give our brains new inputs that make them operate completely differently. We could develop our visual capacities in new ways that will redefine our work and personal lives. And this may well be a way for us to evolve and move forward again, but it's undeniable that something crucial about who we are would be lost in the process.

Conclusion

As I finished my final edits to this book, lives all over the world were upended by the Covid-19 pandemic, and there was an immediate risk to touching. At first, people experimented with new ways to connect. Handshakes were replaced by fist bumps and elbow taps. We started calling each other more often instead of texting. Strangers offered to buy groceries for elderly people and others who were most susceptible to the virus. We went outside for walks. There was enough novelty in this "new normal" that we mostly took it in stride. We started doing all the tactile things that we usually don't have time for, like spending evenings choreographing dances or baking sourdough bread. We celebrated how, by staying apart, we were in it together.

As our isolation dragged on, we went from being a society that feared screens were causing social disconnect to one where the act of being around others was considered irresponsible, and our screens became our lives. We were forced to go inward—to order food from restaurants and get our packages delivered to the doorstep. Work calls, travel plans, conferences, performances, and weddings were canceled, replaced by online meet-ups. Places of worship offered broadcast and drive-through blessings. Women were coached through childbirth on their cellphones. People died in hospital rooms with only a television to keep them company. The lightheartedness went away, and fear crept in.

We had to find more permanent ways to cope with our loss of touch as being close to people, except those who lived with us, was out of the question. Those who were quarantined in thriving families hugged them tighter. They spent more time together without distraction. Others who started off unhappy became even less happy. Cases of domestic violence increased. Those on lockdown alone were especially vulnerable to a dearth of touch, which has affected some more than others, and dating was no longer a solution for those who felt particularly lonely. Trauma doesn't always bring people closer. We each resorted to versions of our typical engagement styles. We became clingy. We isolated ourselves further. Or we sublimated our loneliness in negative or productive ways. Touch has become even more stratified than it already was.

The virus made us all finally reckon with the fact that our institutions can't protect us, and some institutions have actively kept some of us down. We saw how this period affected those who got sick and their families, the essential workers, and those who lost their jobs, including many from underprivileged communities. Some of these communities rose up and took collective action regarding violence directed at them. But what's also been revealed are other subtler inequalities between us, including our access to affection. It may not even be true that we're receiving less touch than before since touch aversion was already on the rise. We do, however, feel its loss more acutely since we have been actively barred from it.

We have the opportunity now to rethink so much of how we live, including how we touch. It is unlikely we will go back to hugging strangers and giving handshakes. It will take longer into courtship for romantic partners to touch each other. Services such as massage therapy and cuddling will struggle, at least for a time. Telework, including telemedicine, could be our future. At the same time, there is some hope that we'll come to appreciate human contact even more. On Twitter, people have posted that they realized now how much they have needed touch and how hard it has been for them without it. They have envisioned a magical time when we'd emerge from our homes and throw cuddle parties and give hugs to strangers. A few have vowed to devote themselves more to cooking and dancing and exercising, to generally slowing down and engaging in physical experiences.

In the short term, we can meet our physiological need for touch simply by caring more for our own bodies. There is certainly solace to be found in being in nature, engaging in hobbies, and enjoying the comforts of home. In the longer term, while considering the need to protect our health, we can also think about how to regain connection to our communities and surroundings and to redesign contemporary life to receive the level of touch we need in our personal relationships, our work, our customs, and our art. We can learn to touch in new ways. But this requires going much further back than the anxiety we have felt during our recent societal shutdown. The way we think about the senses has been flawed for centuries, so it's helpful to examine where we stand to begin with—to look at our most basic of assumptions about the senses and consider where they came from and how we came to ignore our basic need for touch. The reason this matters is that our sense of touch is intimately connected to who we are and what we value.

If we truly want to live with more feeling when this is over, we need to first get in touch with ourselves. We mostly go through life not knowing that our perception is unique to us. Just like cultural anthropologists, we can try to make visible this personalized sensescape by performing a sensory ethnography. Picture the table where you eat breakfast. Is there a predominant image you see? Is that image make-believe or is it a memory? Is it clear or blurry? Do the colors seem like they are filtered through a colored lens or are they realistic?

Do you hear anything—maybe rain pattering on the window or a plane flying in the distance? Are there smells—the toasting of bread or spoiled milk? Do you taste anything—the sweetness of jelly or bitterness of burned toast? Are there flavors that seem like they don't belong, like pickle juice? Is music playing in the background? Is this a tune that actually plays at breakfast time or is this a detail you've superimposed later? Now think about what you feel. What's the temperature? Is there a fan blowing? What is the pattern of the grooves in your cutlery?

What you remember can help you understand your instincts for sensation. Consider not just the most salient of the senses but also how you've called them to mind. Did you relive the experience of using your senses or was it just the words that came up first? Did you have to talk or paint a mental picture in order to conjure them? Did you close your eyes or stare off in space? What was your body language as you were thinking? Did you make any gestures that jogged your memory?

Francis Galton, a Victorian-era psychologist and polymath, first came up with this exercise to understand people's specialized methods for sensing. He found that people from various social classes, sexes, and races recall different details in their mind's eye. Childhood is when we develop our understanding of what the senses mean and how to use them. Being raised in the woods or a bustling city, being required to do housework as a child or encouraged to go to our rooms and study, or making a habit of running through the sprinklers on a hot day versus blasting the air conditioner made us hone in on certain sensory details while repressing others. Certain morals, stories, and art that we were exposed to added to our sensory biases. We weren't born sensing the way we do, even though it might feel like it.

Once we can acknowledge how unobjective we are, then we can try to pinpoint the exact cultural influences that made us this way. A good first place to look is our language. The choice of words in our metaphors and maxims such as "I smell a rat" or "the taste of defeat" reveals deep truths about our sensory beliefs. In the first example, smelling is presented as a hunch rather than a definite, which suggests it's unreliable. In a culture in which smelling is a highly trained sense that's necessary for finding safe food to eat, it might be used differently. In the second example, tasting is presented as a severe or visceral way to experience a negative emotion. That's because we think of it as a sense closely associated with the body, unlike the distance senses such as vision, hearing, and smell.

The types of sensations we use to describe objects also tell us about our preferred methods of perception. For example, our primary sensory association with blood is typically the color red. The taste of blood is also distinct, but that's not what first comes to mind, and the other senses are even less important descriptors. However, in many traditional medicine practices in

Asia, the feeling of the pulse—sharp or slow or strong—is important, so people could be more prone to describe blood using the sense of touch. The Ainu, an indigenous group in Japan and Russia, describes blood primarily by its smell, which they believe can repel spirits. Our instinct is often to use a single sense to describe every object, but this means we're falsely limiting our range of perception.

Next, we can study how the senses appear in art. In the West each of the senses has a distinct medium. This may be because of the way we think of the senses overall, as separate abilities instead of interconnected ways of knowing. But in other cultures, there's an understanding that the senses interact and that it's the cooperative harmony of them working together that creates something that's aesthetically pleasing. For example, a Japanese tea ceremony is meant to impart beauty in multiple forms. A hanging scroll or a flower arrangement is for visual enjoyment. The sound of boiling water purifies the ears. The taste comes from the tea. The pattern of glaze on the pottery and the heat emanating from the bowl are pleasant to touch. Smell is enhanced by the scent of the straw flooring and the steam from the tea.

Mythology and rituals are another source for clues. The tale of Samson and Delilah in the Bible is an illustration of the opposition between vision and touch. Samson was given supernatural strength by God to help deliver Israel from Philistine possession. But then the beautiful prostitute Delilah, who sympathized with the Philistines, seduced Samson. She found the source of his strength, which was his hair, and shaved it off while he slept in her lap. This allowed the Philistines to blind him and enslave him. The story warns that giving in to the carnal pleasures of a woman can make a man lose sight of his greater purpose. The superiority of sight is also present in

Hinduism. In temples, priests bring a small oil lamp around to congregants, who cup their hands over the flame and then reverently touch them to their eyes. Symbolically, they are using their hands to draw the power of the light to the organ most associated with thought as a way of asking for illumination.

Our practices around childrearing are hugely influential because they give us our earliest teachings about the senses. What are the beliefs parents hold—that children should be indulged or disciplined, that they should be held closely or encouraged to be independent? Is spanking permitted? What about the use of a pacifier? At what age is it considered inappropriate for children to sit on laps or be coddled? During children's education, what is emphasized, playing around in a park, building with blocks, listening to stories, or reading quietly? How are the senses used to teach how children should behave? For example, you probably know the phrase "children should be seen and not heard." Sometimes when kids are crying, they're told to toughen up or that they're too soft. Are girls or boys more likely to hear this?

When we start to see where our ideas about the senses come from, then we can rethink them. When we look at objects, we might start thinking about describing them using all of the senses and not just one or two. When out in nature, we might realize how visually based our enjoyment of it is and try to find other forms of appreciation. We might think to pick up a leaf and touch its veins or feel the land under our feet. We could all benefit from detaching from our usual programming and look within, to close our eyes and remind ourselves what it is we feel.

We can learn to use touch purposefully. Practicing yoga or dance or manual therapies can be an important first step to resensitizing the skin and noticing what emotions may lie underneath them. But it doesn't take fancy classes or treatments to reengage touch. It can be as simple as detecting the signals that let us know how to stay balanced when we're walking or that our arms are raised. There's a subtle joy in observing the crunch of cereal in the morning, the cool crispness of hotel bed sheets, and the movement of computer keys.

We can consider taking on hobbies that let us learn a physical or manual skill. We can plant a garden, construct miniature models, or draw sketches out in nature. The act of doing has a way of embedding in our muscle memory and becoming a part of us in a way that book learning does not. Working with our hands requires a focused attention that makes multitasking difficult, which is useful when the buzzes and flashes of our phone have trained us for distraction. These activities can spiral out into metaphors for our lives, which is part of what makes them so rewarding. Through seeing our seeds sprout, we can find regeneration. By climbing a mountain, we can think about the more abstract challenges that we wish to conquer. There are countless ways that our use of the senses affects our worldview.

These are some of the realizations that I made as I reported this book. I noticed how much I desired more physical engagement and yet didn't think I had time for it. More meals than I wanted required dialing and microwaving rather than chopping and kneading, and I had no idea how to fix any of the broken appliances in my home even though I wanted to be self-sufficient. Whenever there was a crisis, someone with the skills was always a phone call or click away. I'd created the kind of life that let me ignore my physical surroundings and stay focused, almost entirely, on what was happening on my devices. I even let my phone override the kinds of basic functions my body was built to do, like deciding whether I'd taken in the right number of calories or gotten enough exercise.

Slowly, that changed. Soon after Kartik and I got married, we adopted a dog. As I took care of my little fifteen-pound poodle-terrier mix, stroking her hair and tossing a ball for her, I felt my relationship with myself changing. It was as if by making these external motions to her, I was also sending them inward, giving myself permission to relax and play. Siggi loves to sleep on my lap on days I work from home, and I've gotten to know her body so well that I can tell her mood, and whether she's feeling sick, based on her breath and the way she carries herself. I tried to begin reading my own emotions the same way, through cues such as my breathing, my shoulders, and the way my skin feels.

I became a more serious baker, which is a wonderful activity for getting out of my head. As I follow the steps, adding wet ingredients to dry, kneading the dough until I felt the right texture under my fingers, and crossing my fingers as it rose in the oven, I can't have lofty thoughts. I'm focused on what's right in front of me. If it comes out wrong, there's no way to intellectualize the mistake. My set of ingredients doesn't have to do what I say. Instead, I have to work to understand them. The small universe that I create with my supplies is a stand-in for how I solve problems in the real world. Even though I'd like people to conform to my wishes, they rarely do, and operating within limits is a lesson I carry from the kitchen counter to the rest of life.

I wrote this chapter, or at least the first draft of it, by hand. It was the first time in years, probably since I was taking finals in college, that I tried to write anything significant with a pen and paper, and it returned me to thinking of my writing as a craft. It was fascinating to recognize again the process by which thoughts could be converted into symbols, which gave it the feeling of a craft, like knitting a sweater or sculpting clay. Slowing down to the pace of my handwriting made me consider each word more carefully. My thoughts flowed more smoothly, and I wasn't constantly deleting and starting over the way I do on a computer. It deepened the experience, if not the words themselves.

On quiet walks, I enjoyed the sensation of my feet pressing against pavement and the way they gave rhythm to my thoughts. I felt the breath moving in and out of my body, the wind hitting my face, and my jeans hugging my skin. I noticed the communication I was having with my car on a rainy day and I felt myself skidding as I tried to brake at a red light. My heart raced as I came to a stop. The sensations I felt in my hands and from my seat all contributed to the notion that my car and I had become one. If we crashed, we were in it together. These small cues, which I hardly noticed before, grounded me in my body and let me know my place in my environment. Even when my attention was pulled in different directions, these were constant reminders of my deeper, personal needs. I used them to think about how I could be more sensitive to myself and others.

The consequences of being out of touch affect us not just personally but societally. Ignoring or devaluing touch is at the core of some our most damaging beliefs. One example is our obsession with thinness. We are trained to use how we look as a marker of our wellness instead of using how we feel. Calorie counting is a virtue, while experiencing the pleasure of good food is a sign of indulgence and weakness. This creates a sensory split within us; we rely too much on vision, never living up to our idealized self-image, and start to ignore how we feel. Our relationship with our bodies can only improve if we listen again to these inner cues of our hunger and enjoyment and let them guide us, which is why the notion of intuitive eating is currently so in vogue.

Pain is one of our most stigmatized medical conditions because there are no biomarkers that can objectively measure it. That makes it easy to dismiss, especially in people who are already the most vulnerable. Historically, medical authorities have perpetuated the notion that black people cannot feel the same degree of pain as white people. Black patients continue to receive less pain relief and management resources than white ones. Similarly, studies show that women are treated for pain less aggressively than men and told it's all in their heads, even though women report pain more often and in higher degrees. We need to change how we respond to people's suffering. When we trust that we all know what we're feeling, it is harder to excuse unequal and cruel treatment of others.

Sealed away in our temperature-controlled homes and offices, we're disconnected from the environment. Even though science is educating us about the observable and calculable changes that our planet is undergoing, what we're missing is our more tangible, intuitive way of understanding our relationship with nature. We don't fully experience the changing of the seasons and can't observe the effects on plants and animals. Our reciprocal relationship with our planet gets lost. Feeling our interdependency is the key to getting people to understand the crisis we're now in, and no number of stories about sinking cities and hurricanes will replace that. Spending time in nature can be an empathetic act, one that we should encourage more in our schools.

Our technology encourages us to embrace the ease of long-distance communication while discounting the value of physical presence, and as we've embraced it our country has become more divided. There's a reason that more people are gravitating toward online dating, where it's possible to project the best version of ourselves and to face rejection without almost any actual hurt. It's easier to communicate with distant old friends through texts than to face true nearness and conflict with a neighbor next door. We'd prefer to exist in our own echo chambers online than to understand those we disagree with. But proximity forces us to deal in the flesh with actual people's feelings, and it asks us for listening, openness, and politeness. By exercising these skills on a regular basis, even with our own families, we become more resilient and tend to feel better about ourselves.

When the threats to our health are less immediate, we should reconsider bans on touch in public life and push for nuanced training programs, from school to the workplace, that can help people recognize how their body language could come across to others and how to think about adjusting. We should be encouraging people to touch with empathy and teaching them how to clearly ask for what they need from each other, particularly because there are more people who are not receiving enough of it in their private lives. People in helping professions such as health care and teaching should be encouraged to see this as an important part of their jobs because they treat people who might not have any other source of affection. Teachers could give lessons on somatic awareness in schools. Concurrently, we need to consistently and appropriately punish people whose behavior is harmful.

Changing attitudes about touch would mean giving more respect to the professionals who, in our touch-starved culture, give us kindness and care that we don't get elsewhere, such as nurses and massage therapists. Too often we talk about the labor they do as being soft and unserious or, in the case of massage, conflate it with sex work, another profession that has been historically stigmatized. (While I was in massage school, I don't know how many times I received comments from friends who thought they were being funny by asking whether I'd learned to give happy endings.) Subconsciously, we continue to view the body as the seat of our emotions that keeps us tethered to a subjective reality. But this is a flawed way of thinking. When bad things happen to us, we want to be infantilized, to have our bodies treated the way they were when we were very young.

Valuing touch is about something much bigger than the sense itself. When we start to pay attention to it, we also observe the dozens of other signals that make us conscious of our inner desires. It's what helps us regain contact with our own interests rather than believing everything we see outside of us. Placing importance on how we feel makes us kinder to others because empathy requires being present to ourselves first. We live in bodies that are most alive when they're open and permeable to what is around us. Touch is a constant affirmation that we exist as selves, separate from our surroundings but connected to them. As a culture, we could aim to live with more feeling, to be as open-palmed as we are clear-eyed.

Notes

1. Dull

1. Rachel Holmes, *Eleanor Marx: A Life* (New York: Bloomsbury, 2015).
2. Sam Shuster, "The Nature and Consequence of Karl Marx's Skin Disease," *British Journal of Dermatology* 158, no. 1 (January 2008): 1–3, https://doi.org/10.1111/j.1365-2133.2007.08282.x.
3. Louis Menand, "Karl Marx, Yesterday and Today," *New Yorker*. October 3, 2016, https://www.newyorker.com/magazine/2016/10/10/karl-marx-yesterday-and-today.
4. Otto Ruhle, *Karl Marx: His Life and Works* (New York: Viking, 1943).
5. Shuster, "Nature and Consequence"; and Philip Jackman, "Marx's Skin Problem," *Globe and Mail*, October 31, 2007, https://www.theglobeandmail.com/news/world/marxs-skin-problems/article696821/.
6. David Howes, *Sensual Relations: Engaging the Senses in Culture and Social Theory* (Ann Arbor: University of Michigan Press, 2010), 230.
7. Howes, *Sensual Relations*, 206.
8. Robert Jütte, *A History of the Senses: From Antiquity to Cyberspace* (Cambridge: Polity, 2005), 10.
9. David Howes, "The Skinscape," *Body & Society* 24, no. 1–2 (2018): 225–39, https://doi.org/10.1177/1357034x18766285.

10. David Howes, "Multisensory Anthropology," *Annual Review of Anthropology* 48 (2019): 17–28, https://www.annualreviews.org/doi/abs/10.1146/annurev-anthro-102218-011324.
11. Howes, *Sensual Relations*, 230.
12. Ian Ritchie, "Fusion of the Faculties: A Study of the Language of the Senses in Hausaland," in *The Varieties of Sensory Experience: A Sourcebook in the Anthropology of the Senses*, ed. David Howes, 192–202 (Toronto: University of Toronto Press, 1991).
13. Sarah Pink, *Doing Sensory Ethnography* (London: Sage, 2009).
14. Peter A. Andersen, "Tactile Traditions: Cultural Differences and Similarities in Haptic Communication," in *The Handbook of Touch: Neuroscience, Behavioral, and Health Perspectives*, ed. Matthew J. Hertenstein and Sandra Jean Weiss, 351–71 (New York: Springer, 2011), 359.
15. Constance Classen, *The Deepest Sense: A Cultural History of Touch* (Urbana: University of Illinois Press, 2012), xii–xiii.
16. Classen, *The Deepest Sense.*
17. Quoted in Sara Danius, "Modernist Fictions of Speed," in *The Book of Touch*, ed. Constance Classen, 412–19 (Oxford: Berg, 2005), 414.
18. Erin Lynch, David Howes, and Martin French, "A Touch of Luck and a 'Real Taste of Vegas': A Sensory Ethnography of the Montreal Casino," *Senses and Society* 15, no. 2 (2020): 192–215, https://www.tandfonline.com/doi/full/10.1080/17458927.2020.1773641.

2. Numb

1. Jonathan Cole, *Pride and a Daily Marathon* (Cambridge, Mass.: MIT Press, 1995).
2. L. A. Goldsmith, "My Organ Is Bigger than Your Organ," *Archives of Dermatology* 126, no. 3 (January 1990): 301–2, https://doi.org/0.1001/archderm.1990.01670270033005.
3. Bruce Goldstein, *Sensation and Perception*, 10th ed. (Boston: Cengage Learning, 2014).
4. Frank R. Wilson, *The Hand: How Its Use Shapes the Brain, Language, and Human Culture* (New York: Vintage, 1999).
5. Jonathan Cole, *Losing Touch: A Man Without His Body* (Oxford: Oxford University Press, 2016).
6. Cole, *Losing Touch.*

7. Sabrina Richards, "Pleasant to the Touch," *Scientist*, September 2012.
8. India Morrison, email to Sushma Subramanian, March 14, 2018.
9. David J. Linden, *Touch: The Science of Hand, Heart, and Mind* (New York: Penguin, 2016).

3. Mushy

1. V. S. Ramachandran and David Brang, "Tactile-Emotion Synesthesia," *Neurocase* 14, no. 5 (December 2008): 390–99, https://doi.org/10.1080/13554790802363746.
2. Pascal Massie, "Touching, Thinking, Being: The Sense of Touch in Aristotle's De Anima and Its Implications." *Minerva: An Internet — Journal of Philosophy* 17 (January 2013): 74–101.
3. Quoted in Massie, "Touching, Thinking, Being," 84.
4. Duncan B. Leitch and Kenneth C. Catania, "Structure, Innervation and Response Properties of Integumentary Sensory Organs in Crocodilians," *Journal of Experimental Biology* 215, no. 23 (July 2012): 4217–30, https://doi.org/10.1242/jeb.076836.
5. Naomi I. Eisenberger, "The Pain of Social Disconnection: Examining the Shared Neural Underpinnings of Physical and Social Pain," *Nature Reviews Neuroscience* 13, no. 6 (March 2012): 421–34, https://doi.org/10.1038/nrn3231.
6. Tristen K. Inagaki, and Naomi I. Eisenberger, "Shared Neural Mechanisms Underlying Social Warmth and Physical Warmth." *Psychological Science* 24, no. 11 (2013): 2272–80. https://doi.org/10.1177/0956797613492773.
7. Joshua M. Ackerman, Christopher C. Nocera, and John A. Bargh, "Incidental Haptic Sensations Influence Social Judgments and Decisions," *Science* 328, no. 5986 (2010): 1712–15, https://doi.org/10.1126/science.1189993.
8. Robert M. Sapolsky, *Behave: The Biology of Humans at Our Best and Worst* (London: Vintage, 2018).
9. Susan Cain, *Quiet: The Power of Introverts in a World That Can't Stop Talking* (New York: Broadway, 2013).
10. Michael J. Banissy and Jamie Ward, "Mirror-Touch Synesthesia Is Linked with Empathy," *Nature Neuroscience* 10, no. 7 (2007): 815–16, https://doi.org/10.1038/nn1926.

11. Nicolas Rothen, and Beat Meier, "Higher Prevalence of Synaesthesia in Art Students," *Perception* 39, no. 5 (2010): 718–20, https://doi.org/10.1068/p6680.
12. Ashley Montagu, *Touching: The Human Significance of the Skin* (New York: Columbia University Press, 1971), 128.
13. Montagu, *Touching*, 5–6.
14. George Lakoff and Mark Johnson, *Metaphors We Live By* (Chicago: University of Chicago Press, 2017).
15. Quoted in Katy Waldman, "Metaphorically Speaking: Our Most Sophisticated Thinking Relies on Bodily Experience," *Slate*, November 24, 2014. Waldman mistakenly attributes this quote to Lakoff and Johnson's *Metaphors We Live By*. It is found in George Lakoff and Mark Johnson, *Philosophy in the Flesh: The Embodied Mind and Its Challenge to Western Thought* (New York: Basic Books, 1999), 4.
16. Pablo Maurette, *The Forgotten Sense* (Chicago: University of Chicago Press, 2018).

4. Untethered

1. Diane Gromala, Xin Tong, Chris Shaw, Ashfaq Amin, Servet Ulas, and Gillian Ramsay, "*Mobius Floe*: An Immersive Virtual Reality Game for Pain Distraction," *Electronic Imaging*, no. 4 (2016): 1–5, https://doi.org/10.2352/issn.2470-1173.2016.4.ervr-413.
2. Diane Gromala, Xin Tong, Chris Shaw, and Weina Jin, "Immersive Virtual Reality as a Non-Pharmacological Analgesic for Pain Management," in *Virtual and Augmented Reality: Concepts, Methodologies, Tools, and Applications*, ed. Information Resources Management Association, 1176–99 (Hershey, Pa.: IGI Global, 2018). https://doi.org/10.4018/978-1-5225-5469-1.ch056.
3. Bernhard Spanlang, Jean-Marie Normand, David Borland, Konstantina Kilteni, Elias Giannopoulos, Ausiàs Pomés, Mar González-Franco, Daniel Perez-Marcos, Jorge Arroyo-Palacios, Xavi Navarro Muncunill, and Mel Slater, "How to Build an Embodiment Lab: Achieving Body Representation Illusions in Virtual Reality," *Frontiers in Robotics and AI* 1 (November 27, 2014), https://doi.org/10.3389/frobt.2014.00009.

4. J. M. S. Pearce, "The Law of Specific Nerve Energies and Sensory Spots," *European Neurology* 54, no. 2 (2005): 115–17, https://doi.org/10.1159/000088647.
5. M. D'Alonzo, A. Mioli, D. Formica, L. Vollero, and G. Di Pino, "Different Level of Virtualization of Sight and Touch Produces the Uncanny Valley of Avatar's Hand Embodiment," *Scientific Reports* 9, no. 1 (2019). https://doi.org/10.1038/s41598-019-55478-z.
6. John Schwenkler, "Do Things Look the Way They Feel?" *Analysis* 73, no. 1 (2012): 86–96, https://doi.org/10.1093/analys/ans137.
7. Nicholas J. Wade, *A Natural History of Vision* (Cambridge, Mass.: MIT Press, 1999), 345.
8. Richard Held, Yuri Ostrovsky, Beatrice de Gelder, Tapan Gandhi, Suma Ganesh, Umang Mathur, and Pawan Sinha, "The Newly Sighted Fail to Match Seen with Felt," *Nature Neuroscience* 14 (2011): 551–553, https://doi.org/10.1038/nn.2795.
9. Denise Grady, "The Vision Thing: Mainly in the Brain," *Discover*, June 1, 1993, https://www.discovermagazine.com/mind/the-vision-thing-mainly-in-the-brain.
10. Quoted in Constance Classen, *The Deepest Sense: A Cultural History of Touch* (Urbana: University of Illinois Press, 2012), 54.
11. Classen, *The Deepest Sense*, 54.
12. Alberto Gallace, Giovanna Soravia, Zaira Cattaneo, Lorimer Moseley, and Giuseppe Vallar, "Temporary Interference over the Posterior Parietal Cortices Disrupts Thermoregulatory Control in Humans," *PLoS ONE* 9, no. 3 (December 2014), https://doi.org/10.1371/journal.pone.0088209.
13. Laura Crucianelli, Nicola K. Metcalf, Katerina Fotopoulou, and Paul M. Jenkinson, "Bodily Pleasure Matters: Velocity of Touch Modulates Body Ownership During the Rubber Hand Illusion," *Frontiers in Psychology* 4 (2013), https://doi.org/10.3389/fpsyg.2013.00703.
14. Laura Crucianelli, Valentina Cardi, Janet Treasure, Paul M. Jenkinson, and Katerina Fotopoulou, "The Perception of Affective Touch in Anorexia Nervosa," *Psychiatry Research* 239 (2016): 72–78, https://doi.org/10.1016/j.psychres.2016.01.078.
15. Rachel Richards, *Hungry for Life: A Memoir Unlocking the Truth Inside an Anorexic Mind* (self-pub., 2016).

16. G. Lorimer Moseley, Timothy J. Parsons, and Charles Spence, "Visual Distortion of a Limb Modulates the Pain and Swelling Evoked by Movement," *Current Biology* 18, no. 22 (2008), https://doi.org/10.1016/j.cub.2008.09.031.
17. Matt Haber, "A Trip to Camp to Break an Addiction," *New York Times*, July 5, 2013, https://www.nytimes.com/2013/07/07/fashion/a-trip-to-camp-to-break-a-tech-addiction.html.
18. Andrew Sullivan, "I Used to Be a Human Being," *New York Magazine*, September 19, 2016, https://nymag.com/intelligencer/2016/09/andrew-sullivan-my-distraction-sickness-and-yours.html.
19. Richard Kearney, "Losing Our Touch," *New York Times*, August 30, 2014, https://opinionator.blogs.nytimes.com/2014/08/30/losing-our-touch/.
20. Richard Kearney and Brian Treanor, *Carnal Hermeneutics* (New York: Fordham University Press, 2015).

5. Softening

1. Stanley E. Jones and Brandi C. Brown, "Touch Attitudes and Behaviors, Recollections of Early Childhood Touch, and Social Self-Confidence," *Journal of Nonverbal Behavior* 20, no. 3 (1996): 147–63, https://doi.org/10.1007/bf02281953.
2. Peter A. Andersen and Karen Kuish Sull, "Out of Touch, Out of Reach: Tactile Predispositions as Predictors of Interpersonal Distance," *Western Journal of Speech Communication* 49, no. 1 (1985): 57–72, https://doi.org/10.1080/10570318509374181.
3. William J. Chopik, Robin S. Edelstein, Sari M. van Anders, Britney M. Wardecker, Emily L. Shipman, and Chelsea R. Samples-Steele, "Too Close for Comfort? Adult Attachment and Cuddling in Romantic and Parent–Child Relationships," *Personality and Individual Differences* 69 (2014): 212–16, https://doi.org/10.1016/j.paid.2014.05.035.
4. Deborah Blum, *Love at Goon Park: Harry Harlow and the Science of Affection* (New York: Basic Books, 2011).
5. Jean O'Malley Halley, *Boundaries of Touch: Parenting and Adult–Child Intimacy* (Urbana: University of Illinois Press, 2009).
6. Blum, *Love at Goon Park*, 37.

7. Maria Konnikova, "The Power of Touch," *New Yorker*, March 4, 2015, https://www.newyorker.com/science/maria-konnikova/power-touch.
8. Tiffany Field, *Touch* (Cambridge, Mass.: MIT Press, 2001).
9. Kory Floyd, "Relational and Health Correlates of Affection Deprivation," *Western Journal of Communication* 78, no. 4 (2014): 383–403, https://doi.org/10.1080/10570314.2014.927071.
10. Lana Bestbier and Tim I. Williams, "The Immediate Effects of Deep Pressure on Young People with Autism and Severe Intellectual Difficulties: Demonstrating Individual Differences," *Occupational Therapy International* 2017 (2017): 1–7, https://doi.org/10.1155/2017/7534972.
11. Susan Bauer, *The Embodied Teen: A Somatic Curriculum for Teaching Body-Mind Awareness, Kinesthetic Intelligence, and Social and Emotional Skills* (Berkeley, Calif.: North Atlantic, 2018).
12. Lorraine Green, "The Trouble with Touch? New Insights and Observations on Touch for Social Work and Social Care." *British Journal of Social Work* 47, no. 3 (April 2017): 773–92, https://doi.org/10.1093/bjsw/bcw071.
13. Bessel Van der Kolk, *The Body Keeps the Score: Brain, Mind and Body in the Healing of Trauma* (New York: Penguin, 2015).
14. Michael W. Kraus, Casey Huang, and Dacher Keltner, "Tactile Communication, Cooperation, and Performance: An Ethological Study of the NBA," *Emotion* 10, no. 5 (2010): 745–49, https://doi.org/10.1037/a0019382.
15. Tiffany Field, "American Adolescents Touch Each Other Less and Are More Aggressive Toward Their Peers as Compared with French Adolescents," *Adolescence* 34, no. 136 (Winter 1999): 752–58.
16. James W. Prescott, "Body Pleasure and the Origins of Violence," *Bulletin of the Atomic Scientists* 31, no. 9 (1975): 10–20, https://doi.org/10.1080/00963402.1975.11458292.

6. Boundaries

1. Jean O'Malley Halley, *Boundaries of Touch: Parenting & Adult-Child Intimacy* (Chicago: University of Illinois, 2009).
2. Russell Clark and Elaine Hatfield, "Gender Differences in Receptivity to Sexual Offers," *Journal of Psychology & Human Sexuality* 2, no. 1 (July 1989): 39–55, https://doi.org/10.1300/j056v02n01_04.

3. Halley, *Boundaries of Touch*.
4. Reva B. Siegel, "Introduction: A Short History of Sexual Harassment," in *Directions in Sexual Harassment Law*, ed. Catharine A. MacKinnon and Reva B. Siegel, 1–28 (New Haven, Conn.: Yale University Press, October 2003). https://doi.org/10.12987/yale/9780300098006.003.0001.
5. Ofer Zur, "Touch in Therapy," in *Boundaries in Psychotherapy: Ethical and Clinical Explorations*, 167–85 (Washington, D.C.: American Psychological Association, 2007), https://doi.org/10.1037/11563-010.
6. Joseph L. Daly, "'Gray Touch': Professional Issues in the Uncertain Zone Between 'Good Touch' and 'Bad Touch.'" *Marquette Elder's Advisor* 11, no. 2: 223–79.
7. Michelle Obama, *Becoming* (New York: Crown, 2018), 318.
8. Dan Pens, "Skin Blind," in *Prison Masculinities*, ed. Donald F. Sabo, Terry Allen Kupers, and Willie James London, 150–52 (Philadelphia: Temple University Press, 2001).
9. Pens, "Skin Blind," 150, 151, 152.
10. Joanna Bourke, *Dismembering the Male: Men's Bodies, Britain and the Great War* (London: Reaktion, 2009).
11. Eric Anderson, *21st Century Jocks: Sporting Men and Contemporary Heterosexuality* (Basingstoke, Hampshire: Palgrave Macmillan, 2015).
12. Jean M. Twenge, Ryne A. Sherman, and Brooke E. Wells, "Sexual Inactivity During Young Adulthood Is More Common Among U.S. Millennials and IGen: Age, Period, and Cohort Effects on Having No Sexual Partners After Age 18," *Archives of Sexual Behavior* 46, no. 2 (January 2016): 433–40, https://doi.org/10.1007/s10508-016-0798-z.
13. Kate Julian, "Why Are Young People Having So Little Sex?" *Atlantic*, December 2018, https://www.theatlantic.com/magazine/archive/2018/12/the-sex-recession/573949/.
14. Hope Reese, "Americans Are Having Less Sex, but Is That a Problem?" *Greater Good Magazine*, February 18, 2019, https://greatergood.berkeley.edu/article/item/americans_are_having_less_sex_but_is_that_a_problem.
15. William J. Chopik, Robin S. Edelstein, Sari M. van Anders, Britney M. Wardecker, Emily L. Shipman, and Chelsea R. Samples-Steele, "Too Close for Comfort? Adult Attachment and Cuddling in

Romantic and Parent–Child Relationships," *Personality and Individual Differences* 69 (2014): 212–16, https://doi.org/10.1016/j.paid.2014.05.035.

16. Jill Lepore, "The History of Loneliness," *New Yorker*, April 6, 2020, https://www.newyorker.com/magazine/2020/04/06/the-history-of-loneliness.
17. Jena McGregor, "This Former Surgeon General Says There's a 'Loneliness Epidemic' and Work Is Partly to Blame," *Washington Post*, October 4, 2017.
18. J. Richard Udry, "The Effect of the Great Blackout of 1965 on Births in New York City," *Demography* 7, no. 3 (1970): 325, https://doi.org/10.2307/2060151.
19. Joseph Lee Rodgers, Craig A. St. John, and Ronnie Coleman, "Did Fertility Go Up After the Oklahoma City Bombing? An Analysis of Births in Metropolitan Counties in Oklahoma, 1990–1999," *Demography* 42, no. 4 (2005): 675–92, https://doi.org/10.1353/dem.2005.0034.
20. Nellie Bowles, "Human Contact Is Now a Luxury Good," *New York Times*, March 23, 2019, https://www.nytimes.com/2019/03/23/sunday-review/human-contact-luxury-screens.html.
21. Courtney Maum, *Touch* (New York: G. P. Putnam's Sons, 2017).

7. Slick

1. Anne Saint-Eve, Enkelejda Paçi Kora, and Nathalie Martin, "Impact of the Olfactory Quality and Chemical Complexity of the Flavouring Agent on the Texture of Low Fat Stirred Yogurts Assessed by Three Different Sensory Methodologies," *Food Quality and Preference* 15, no. 7–8 (2004): 655–68, https://doi.org/10.1016/j.foodqual.2003.09.002.
2. Francine Lenfant, Christophe Hartmann, Brigitte Watzke, Oliver Breton, Chrystel Loret, and Nathalie Martin, "Impact of the Shape on Sensory Properties of Individual Dark Chocolate Pieces," *LWT—Food Science and Technology* 51, no. 2 (2013): 545–52, https://doi.org/10.1016/j.lwt.2012.11.001.
3. Ingemar Pettersson, "Mechanical Tasting: Sensory Science and the Flavorization of Food Production," *The Senses and Society* 12, no. 3 (February 2017): 301–16, https://doi.org/10.1080/17458927.2017.1376440.

4. Lorraine Daston and Peter Galison, *Objectivity* (New York: Zone Books, 2007).
5. David Howes and Constance Classen, *Ways of Sensing: Understanding the Senses in Society* (London: Routledge, 2014).
6. Dominik Wujastyk, *The Roots of Ayurveda: Selections from Sanskrit Medical Writings* (London: Penguin, 1998), 268.
7. Axel Munthe, *The Story of San Michele* (Hamburg: Albatross, 1935), 49.
8. Chia Longman, "Women's Circles and the Rise of the New Feminine: Reclaiming Sisterhood, Spirituality, and Wellbeing," *Religions* 9, no. 1 (January 2018): 9, https://doi.org/10.3390/rel9010009.
9. Joseph E. Davis, "The Commodification of Self," *Hedgehog Review: Critical Reflections on Contemporary Culture*, Summer 2003, https://hedgehogreview.com/issues/the-commodification-of-everything/articles/the-commodification-of-self.
10. Julia Emberley, *Venus and Furs: Cultural Politics of Fur* (London: Cornell University, 1998).

8. Haptics

1. David Parisi, *Archaeologies of Touch: Interfacing with Haptics from Electricity to Computing* (Minneapolis: University of Minnesota, 2018).
2. Molly McHugh, "Yes, There Is a Difference Between 3D Touch and Force Touch," *Wired*, September 9, 2015, https://www.wired.com/2015/09/what-is-the-difference-between-apple-iphone-3d-touch-and-force-touch/.
3. Joe Mullenbach, Craig Shultz, Anne Marie Piper, Michael Peshkin, and J. Edward Colgate, "Surface Haptic Interactions with a TPad Tablet," in *UIST 13 Adjunct: Proceedings of the Adjunct Publication of the 26th Annual ACM Symposium on User Interface Software and Technology* (New York: Association for Computing Machinery, 2013), https://doi.org/10.1145/2508468.2514929.
4. David J. Linden, *Touch: The Science of Hand, Heart, and Mind* (New York: Penguin, 2016).
5. Amy M. Green, *Storytelling in Video Games: The Art of the Digital Narrative* (Jefferson, N.C.: McFarland, 2018); and Kazu Masudul Alam, Abu Saleh Md Mahfujur Rahman, and Abdulmotaleb El Saddik,

"HE-Book: A Prototype Haptic Interface for Immersive e-Book Reading Experience," in *2011 IEEE World Haptics Conference*, Istanbul (2011), 367–71, https://doi.org/10.1109/whc.2011.5945514.

6. Ferris Jabr, "The Reading Brain in the Digital Age: The Science of Paper Versus Screens." *Scientific American*, April 11, 2013, https://www.scientificamerican.com/article/reading-paper-screens/.
7. Maryanne Wolf, *Proust and the Squid: The Story and Science of the Reading Brain* (New York: Harper, 2007).
8. Anne Mangen, Bentie R. Walgermo, and Kolbjørn Brønnick, "Reading Linear Texts on Paper Versus Computer Screen: Effects on Reading Comprehension," *International Journal of Educational Research* 58 (2013): 61–68, https://doi.org/10.1016/j.ijer.2012.12.002.
9. Pam A. Mueller and Daniel M Oppenheimer, "The Pen Is Mightier Than the Keyboard," *Psychological Science* 25, no. 6 (2014): 1159–68, https://doi.org/10.1177/0956797614524581.
10. Jane Austen, *Persuasion* (London: Penguin, 1991), 56–57.
11. Patricia Highsmith, *The Price of Salt* (Mineola: Dover Publications, 2017), 38.
12. Quoted in Joanna Briscoe, "Look, Don't Touch: What Great Literature Can Teach Us About Love with No Contact," *Guardian*, May 22, 2020.
13. Alberto Gallace and Charles Spence, *In Touch with the Future: The Sense of Touch from Cognitive Neuroscience to Virtual Reality* (Oxford: Oxford University Press, 2014).

9. Tactful

1. MIT Technology Review, "Restoring a Sense of Touch in Amputees," Youtube video, 2:54, March 21, 2014, https://www.youtube.com/watch?v=075lJjJDonA.
2. Charles Darwin, *Descent of Man and Selection in Relation to Sex* (New York: D. Appleton and Co., 1989), 135.
3. Frank R. Wilson, *The Hand: How Its Use Shapes the Brain, Language, and Human Culture* (New York: Vintage, 1999).
4. Michael Wong, Vishi Gnanakumaran, and Daniel Goldreich, "Tactile Spatial Acuity Enhancement in Blindness: Evidence for

Experience-Dependent Mechanisms," *Journal of Neuroscience* 31, no. 19 (November 2011): 7028–37, https://doi.org/10.1523/jneurosci.6461-10.2011.

5. Robin Dunbar, *Grooming, Gossip and the Evolution of Language* (London: Faber and Faber, 2004).
6. Dunbar, *Grooming, Gossip*, 147, 148.
7. Frank R. Wilson, *The Hand: How Its Use Shapes the Brain, Language, and Human Culture* (New York: Vintage, 1999).
8. Wilson, *The Hand*, 219.

Bibliography

Acciavatti, Anthony. "Ingestion: The Psychorheology of Everyday Life." *Cabinet* 48 (2012–13): 12–16.

Ackerman, Diane. *A Natural History of the Senses*. New York: Random House, 1991.

Ackerman, Joshua M., Christopher C. Nocera, and John A. Bargh. "Incidental Haptic Sensations Influence Social Judgments and Decisions." *Science* 328, no. 5986 (2010): 1712–15. https://doi.org/10.1126/science.1189993.

Alam, Kazi Masudul, Abu Saleh Md Mahfujur Rahman, and Abdulmotaleb El Saddik. "HE-Book: A Prototype Haptic Interface for Immersive e-Book Reading Experience." In *2011 IEEE World Haptics Conference*, Istanbul (2011), 367–71. https://doi.org/10.1109/whc.2011.5945514.

Andersen, Peter A. *Nonverbal Communication: Forms and Functions*. Long Grove, Ill.: Waveland, 2008.

——. "Tactile Traditions: Cultural Differences and Similarities in Haptic Communication." In *The Handbook of Touch: Neuroscience, Behavioral, and Health Perspectives*, ed. Matthew J. Hertenstein and Sandra Jean Weiss, 351–71. New York: Springer, 2011.

Andersen, Peter A., and Karen Kuish Sull. "Out of Touch, Out of Reach: Tactile Predispositions as Predictors of Interpersonal Distance." *Western Journal of Speech Communication* 49, no. 1 (1985): 57–72. https://doi.org/10.1080/10570318509374181.

Andersen, Ross. "A Journey into the Animal Mind." *Atlantic*, March 2019. https://www.theatlantic.com/magazine/archive/2019/03/what-the-crow-knows/580726/.

Anderson, Eric. *21st Century Jocks: Sporting Men and Contemporary Heterosexuality*. Basingstoke, Hampshire: Palgrave Macmillan, 2015.

Antony, Mary Grace. "Thats a Stretch: Reconstructing, Rearticulating, and Commodifying Yoga." *Frontiers in Communication* 3 (2018). https://doi.org/10.3389/fcomm.2018.00047.

Austen, Jane. *Persuasion*. London: Penguin, 1991.

Banissy, Michael J., and Jamie Ward. "Mirror-Touch Synesthesia Is Linked with Empathy." *Nature Neuroscience* 10, no. 7 (2007): 815–16. https://doi.org/10.1038/nn1926.

Bargh, John A., and Erin L. Williams. "The Automaticity of Social Life." *Current Directions in Psychological Science* 15, no. 1 (2006): 1–4. https://doi.org/10.1111/j.0963-7214.2006.00395.x.

Bauer, Susan. *The Embodied Teen: A Somatic Curriculum for Teaching Body-Mind Awareness, Kinesthetic Intelligence, and Social and Emotional Skills*. Berkeley, Calif.: North Atlantic, 2018.

Berila, Beth, Melanie Klein, and Chelsea Jackson Roberts. *Yoga, the Body, and Embodied Social Change: An Intersectional Feminist Analysis*. Lanham, Md.: Lexington, 2016.

Bestbier, Lana, and Tim I. Williams. "The Immediate Effects of Deep Pressure on Young People with Autism and Severe Intellectual Difficulties: Demonstrating Individual Differences." *Occupational Therapy International*, January 9, 2017, pp. 1–7. https://doi.org/10.1155/2017/7534972.

Blum, Deborah. *Love at Goon Park: Harry Harlow and the Science of Affection*. New York: Basic Books, 2011.

Bourke, Joanna. *Dismembering the Male: Men's Bodies, Britain and the Great War*. London: Reaktion, 2009.

Cain, Susan. *Quiet: The Power of Introverts in a World That Can't Stop Talking*. New York: Broadway, 2013.

Chopik, William J., Robin S. Edelstein, Sari M. van Anders, Britney M. Wardecker, Emily L. Shipman, and Chelsea R. Samples-Steele. "Too Close for Comfort? Adult Attachment and Cuddling in Romantic and Parent–Child Relationships." *Personality and Individual Differences* 69 (2014): 212–16. https://doi.org/10.1016/j.paid.2014.05.035.

Clark, Russell, and Elaine Hatfield. "Gender Differences in Receptivity to Sexual Offers." *Journal of Psychology & Human Sexuality* 2, no. 1 (July 1989): 39–55. https://doi.org/10.1300/j056v02n01_04.

Classen, Constance, ed. *The Book of Touch*. Oxford: Berg, 2005.

——. *The Deepest Sense: A Cultural History of Touch*. Urbana: University of Illinois Press, 2012.

Cole, Jonathan. *Losing Touch: A Man Without His Body*. Oxford: Oxford University Press, 2016.

——. *Pride and a Daily Marathon*. Cambridge, Mass.: MIT Press, 1995.

Craig, A. D. "How Do You Feel? Interoception: The Sense of the Physiological Condition of the Body." *Nature Reviews Neuroscience* 3, no. 8 (2002): 655–66. https://doi.org/10.1038/nrn894.

Craig, A. D., and L. S. Sorkin. "Pain and Analgesia." *Encyclopedia of Life Sciences*, 2005. https://doi.org/10.1038/npg.els.0004062.

Crucianelli, Laura, Valentina Cardi, Janet Treasure, Paul M. Jenkinson, and Katerina Fotopoulou. "The Perception of Affective Touch in Anorexia Nervosa." *Psychiatry Research* 239 (2016): 72–78. https://doi.org/10.1016/j.psychres.2016.01.078.

Crucianelli, Laura, Nicola K. Metcalf, Katerina Fotopoulou, and Paul M Jenkinson. "Bodily Pleasure Matters: Velocity of Touch Modulates Body Ownership During the Rubber Hand Illusion." *Frontiers in Psychology* 4 (2013). https://doi.org/10.3389/fpsyg.2013.00703.

D'Alonzo, M., A. Mioli, D. Formica, L. Vollero, and G. Di Pino. "Different Level of Virtualization of Sight and Touch Produces the Uncanny Valley of Avatar's Hand Embodiment." *Scientific Reports* 9, no. 1 (2019). https://doi.org/10.1038/s41598-019-55478-z.

Dacher, Keltner. "Hands on Research: The Science of Touch." *Greater Good Magazine*, September 29, 2010. https://greatergood.berkeley.edu/article/item/hands_on_research.

Daly, Joseph L. "'Gray Touch': Professional Issues in the Uncertain Zone Between 'Good Touch' and 'Bad Touch.'" *Marquette Elder's Advisor* 11, no. 2: 223–79.

Danielsson, Louise, and Susanne Rosberg. "Opening Toward Life: Experiences of Basic Body Awareness Therapy in Persons with Major Depression." *International Journal of Qualitative Studies on Health and Well-Being* 10, no. 1 (2015): 27069. https://doi.org/10.3402/qhw.v10.27069.

Danius, Sara. "Modernist Fictions of Speed." In *The Book of Touch*, ed. Constance Classen, 412–19. Oxford: Berg, 2005.

Darwin, Charles. *Descent of Man and Selection in Relation to Sex*. New York: D. Appleton, 1989.

Daston, Lorraine, and Peter Galison. *Objectivity*. New York: Zone Books, 2007.

Davis, Joseph E. "The Commodification of Self." *Hedgehog Review: Critical Reflections on Contemporary Culture*, Summer 2003. https://hedgehogreview.com/issues/the-commodification-of-everything/articles/the-commodification-of-self.

Drazin, Adam, and Susanne Kueüchler. *The Social Life of Materials: Studies in Materials and Society*. London: Bloomsbury Academic, 2015.

Dunbar, Robin. *Grooming, Gossip and the Evolution of Language*. London: Faber and Faber, 2004.

Eisenberger, Naomi I. "The Pain of Social Disconnection: Examining the Shared Neural Underpinnings of Physical and Social Pain." *Nature Reviews Neuroscience* 13, no. 6 (March 2012): 421–34. https://doi.org/10.1038/nrn3231.

Eisenberger, Naomi I., Johanna M. Jarcho, Matthew D. Lieberman, and Bruce D. Naliboff. "An Experimental Study of Shared Sensitivity to Physical Pain and Social Rejection." *Pain* 126, no. 1 (2006): 132–38. https://doi.org/10.1016/j.pain.2006.06.024.

Emberley, Julia. *Venus and Furs: Cultural Politics of Fur*. Ithaca, N.Y.: Cornell University Press, 1998.

Field, Tiffany. "American Adolescents Touch Each Other Less and Are More Aggressive Toward Their Peers as Compared with French Adolescents." *Adolescence* 34, no. 136 (Winter 1999): 752–58.

——. *Touch*. Cambridge, Mass.: MIT Press, 2001.

——. "Touch Deprivation and Aggression Against Self Among Adolescents." *Developmental Psychobiology of Aggression*, June 2005, 117–40. https://doi.org/10.1017/cbo9780511499883.007.

Floyd, Kory. "Relational and Health Correlates of Affection Deprivation." *Western Journal of Communication* 78, no. 4 (2014): 383–403. https://doi.org/10.1080/10570314.2014.927071.

Fulkerson, Matthew. *The First Sense: A Philosophical Study of Human Touch*. Cambridge, Mass.: MIT Press, 2014.

Gallace, Alberto, and Charles Spence. *In Touch with the Future: The Sense of Touch from Cognitive Neuroscience to Virtual Reality.* Oxford: Oxford University Press, 2014.

Gallace, Alberto, Giovanna Soravia, Zaira Cattaneo, Lorimer Moseley, and Giuseppe Vallar. "Temporary Interference over the Posterior Parietal Cortices Disrupts Thermoregulatory Control in Humans." *PLoS ONE* 9, no. 3 (December 2014). https://doi.org/10.1371/journal.pone.0088209.

Goldin-Meadow, Susan. "Talking and Thinking with Our Hands." *Current Directions in Psychological Science* 15, no. 1 (2006): 34–39. https://doi.org/10.1111/j.0963-7214.2006.00402.x.

Goldsmith, L. A. "My Organ Is Bigger than Your Organ." *Archives of Dermatology* 126, no. 3 (January 1990): 301–2. https://doi.org/10.1001/archderm.1990.01670270033005.

Goldstein, Bruce. *Sensation and Perception.* 10th ed. Boston: Cengage Learning, 2014.

Grady, Denise. "The Vision Thing: Mainly in the Brain." *Discover*, June 1, 1993. https://www.discovermagazine.com/mind/the-vision-thing-mainly-in-the-brain.

Green, Amy M. *Storytelling in Video Games: The Art of the Digital Narrative.* Jefferson, NC: McFarland, 2018.

Green, Lorraine. "The Trouble with Touch? New Insights and Observations on Touch for Social Work and Social Care." *British Journal of Social Work*, 2016. https://doi.org/10.1093/bjsw/bcw071.

Gromala, Diane, Xin Tong, Chris Shaw, Ashfaq Amin, Servet Ulas, and Gillian Ramsay. "*Mobius Floe*: An Immersive Virtual Reality Game for Pain Distraction." *Electronic Imaging*, no. 4 (2016): 1–5. https://doi.org/10.2352/issn.2470-1173.2016.4.ervr-413.

Gromala, Diane, Xin Tong, Chris Shaw, and Weina Jin. "Immersive Virtual Reality as a Non-Pharmacological Analgesic for Pain Management." In *Virtual and Augmented Reality: Concepts, Methodologies, Tools, and Applications*, ed. Information Resources Management Association, 1176–99. Hershey, Pa.: IGI Global, 2018. https://doi.org/10.4018/978-1-5225-5469-1.ch056.

Halley, Jean. *Boundaries of Touch: Parenting and Adult–Child Intimacy.* Urbana: University of Illinois, 2009.

Hamblin, James. "Can We Touch?" *Atlantic*, April 10, 2019. https://www.theatlantic.com/health/archive/2019/04/on-touch/586588/.

Hayward, Vincent. "A Brief Taxonomy of Tactile Illusions and Demonstrations That Can Be Done in a Hardware Store." *Brain Research Bulletin* 75, no. 6 (2008): 742–52. https://doi.org/10.1016/j.brainresbull.2008.01.008.

Held, Richard, Yuri Ostrovsky, Beatrice de Gelder, Tapan Gandhi, Suma Ganesh, Umang Mathur, and Pawan Sinha. "The Newly Sighted Fail to Match Seen with Felt." *Nature Neuroscience* 14 (2011): 551–53. https://doi.org/10.1038/nn.2795.

Hertenstein, Matthew J., and Sandra Jean Weiss. *The Handbook of Touch: Neuroscience, Behavioral, and Health Perspectives*. New York: Springer, 2011.

Hess, Samantha. *Touch: The Power of Human Connection*. Portland, Ore.: Fulcrum Solutions LLC, 2014.

Highsmith, Patricia. *The Price of Salt*. Mineola, N.Y.: Dover, 2017.

Holmes, Rachel. *Eleanor Marx: A Life*. New York: Bloomsbury, 2015.

Howes, David. *A Cultural History of the Senses in the Modern Age*. London: Bloomsbury Academic, 2019.

——. "Multisensory Anthropology." *Annual Review of Anthropology* 48 (2019): 17–28. https://www.annualreviews.org/doi/abs/10.1146/annurev-anthro-102218-011324.

——. *Senses and Sensation: Critical and Primary Sources*. London: Bloomsbury, 2018.

——. *Sensual Relations: Engaging the Senses in Culture and Social Theory*. Ann Arbor: University of Michigan Press, 2010.

——. "The Skinscape." *Body & Society* 24, no. 1–2 (2018): 225–39. https://doi.org/10.1177/1357034x18766285.

——. The Varieties of Sensory Experience: A Sourcebook in the Anthropology of the Senses. Toronto: University of Toronto Press, 1991.

Howes, David, and Classen, Constance. *Ways of Sensing: Understanding the Senses in Society*. London: Routledge, 2014.

Hutson, Matthew. "Here's What the Future of Haptic Technology Looks (Or Rather, Feels) Like." *Smithsonian Magazine*, December 28, 2018. https://www.smithsonianmag.com/innovation/heres-what-future-haptic-technology-looks-or-rather-feels-180971097/.

Ibson, John. *Picturing Men: A Century of Male Relationships in Everyday American Photography*. Chicago: University of Chicago Press, 2006.

Inagaki, Tristen K., and Naomi I. Eisenberger. "Shared Neural Mechanisms Underlying Social Warmth and Physical Warmth." Psychological Science 24, no. 11 (2013): 2272–80. https://doi.org/10.1177/0956797613492773.

Jablonski, Nina G. *Skin: A Natural History*. Berkeley: University of California Press, 2013.

Jabr, Ferris. "The Reading Brain in the Digital Age: The Science of Paper Versus Screens." *Scientific American*, April 11, 2013. https://www.scientificamerican.com/article/reading-paper-screens/.

Jenkinson, Paul. "Self-Reported Interoceptive Deficits in Eating Disorders: A Meta-Analysis of Studies Using the Eating Disorder Inventory." *Journal of Psychosomatic Research* 110 (July 2018): 38–45.

Jones, Stanley E., and Brandi C. Brown. "Touch Attitudes and Behaviors, Recollections of Early Childhood Touch, and Social Self-Confidence." *Journal of Nonverbal Behavior* 20, no. 3 (1996): 147–63. https://doi.org/10.1007/bf02281953.

Julian, Kate. "Why Are Young People Having So Little Sex?" *Atlantic*, December 2018. https://www.theatlantic.com/magazine/archive/2018/12/the-sex-recession/573949/.

Jütte Robert. *A History of the Senses: from Antiquity to Cyberspace*. Cambridge: Polity, 2005.

Kearney, Richard, and Briand Treanor. *Carnal Hermeneutics*. New York: Fordham University Press, 2015.

Konnikova, Maria. "The Power of Touch." *New Yorker*, March 4, 2015. https://www.newyorker.com/science/maria-konnikova/power-touch.

Kraus, Michael W., Casey Huang, and Dacher Keltner. "Tactile Communication, Cooperation, and Performance: An Ethological Study of the NBA." *Emotion* 10, no. 5 (2010): 745–49. https://doi.org/10.1037/a0019382.

Kuhtz-Buschbeck, Johann P., Jochen Schaefer, and Nicolaus Wilder. "Mechanosensitivity: From Aristotle's Sense of Touch to Cardiac Mechano-Electric Coupling." *Progress in Biophysics and Molecular Biology* 130 (2017): 126–31. https://doi.org/10.1016/j.pbiomolbio.2017.05.001.

Lakoff, George, and Mark Johnson. *Metaphors We Live By*. Chicago: University of Chicago Press, 2017.

———. *Philosophy in the Flesh: The Embodied Mind and Its Challenge to Western Thought*. New York: Basic Books, 1999.

Leitch, Duncan B., and Kenneth C. Catania. "Structure, Innervation and Response Properties of Integumentary Sensory Organs in Crocodilians." *Journal of Experimental Biology* 215, no. 23 (July 2012): 4217–30. https://doi.org/10.1242/jeb.076836.

Lenfant, Francine, Christophe Hartmann, Brigitte Watzke, Olivier Breton, Chrystel Loret, and Nathalie Martin. "Impact of the Shape on Sensory Properties of Individual Dark Chocolate Pieces." *LWT—Food Science and Technology* 51, no. 2 (2013): 545–52. https://doi.org/10.1016/j.lwt.2012.11.001.

Lepore, Jill. "The History of Loneliness." *New Yorker*, April 6, 2020. https://www.newyorker.com/magazine/2020/04/06/the-history-of-loneliness.

Linden, David J. *Touch: The Science of Hand, Heart, and Mind*. New York: Penguin, 2016.

Longman, Chia. "Women's Circles and the Rise of the New Feminine: Reclaiming Sisterhood, Spirituality, and Wellbeing." *Religions* 9, no. 1 (January 2018): 9. https://doi.org/10.3390/rel9010009.

Lundborg, Göran. *The Hand and the Brain: from Lucy's Thumb to the Thought-Controlled Robotic Hand*. London: Springer, 2014.

Lynch, Erin, David Howes, and Martin French. "A Touch of Luck and a 'Real Taste of Vegas': A Sensory Ethnography of the Montreal Casino." *Senses and Society* 15, no. 2 (2020): 192–215. https://www.tandfonline.com/doi/full/10.1080/17458927.2020.1773641.

Mangen, Anne, Bente R. Walgermo, and Kolbjørn Brønnick. "Reading Linear Texts on Paper versus Computer Screen: Effects on Reading Comprehension." *International Journal of Educational Research* 58 (2013): 61–68. https://doi.org/10.1016/j.ijer.2012.12.002.

Marx, Karl, and Eugene Kamenka. *The Portable Karl Marx*. New York: Viking, 1983.

Massie, Pascal. "Touching, Thinking, Being: The Sense of Touch in Aristotle's De Anima and Its Implications." *Minerva: An Internet Journal of Philosophy* 17 (January 2013): 74–101.

Maum, Courtney. *Touch*. New York: G. P. Putnam's Sons, 2017.

Maurette, Pablo. *The Forgotten Sense*. Chicago: University of Chicago Press, 2018.

——. *Meditations on Touch*. Chicago: University of Chicago Press, 2018.

Menand, Louis. "Karl Marx, Yesterday and Today." *New Yorker*, October 3, 2016. https://www.newyorker.com/magazine/2016/10/10/karl-marx-yesterday-and-today.

MIT Technology Review. "Restoring a Sense of Touch in Amputees." Youtube video, 2:54. March 21, 2014. https://www.youtube.com/watch?v=075lJjJDonA.

Montagu, Ashley. *Touching: The Human Significance of the Skin*. New York: Columbia University Press, 1971.

Moseley, G. Lorimer, Timothy J. Parsons, and Charles Spence. "Visual Distortion of a Limb Modulates the Pain and Swelling Evoked by Movement." *Current Biology* 18, no. 22 (2008). https://doi.org/10.1016/j.cub.2008.09.031.

Mueller, Pam A., and Daniel M. Oppenheimer. "The Pen Is Mightier Than the Keyboard." *Psychological Science* 25, no. 6 (2014): 1159–68. https://doi.org/10.1177/0956797614524581.

Mullenbach, Joe, Craig Shultz, Anne Marie Piper, Michael Peshkin, and J. Edward Colgate. "Surface Haptic Interactions with a TPad Tablet." *UIST '13 Adjunct: Proceedings of the Adjunct Publication of the 26th Annual ACM Symposium on User Interface Software and Technology*. New York: Association for Computing Machinery, 2013. https://doi.org/10.1145/2508468.2514929.

Munthe, Axel. *The Story of San Michele*. Hamburg: Albatross, 1935.

Obama, Michelle. *Becoming*. New York: Crown, 2018.

Parisi, David. *Archaeologies of Touch: Interfacing with Haptics from Electricity to Computing*. Minneapolis: University of Minnesota, 2018.

Paterson, Mark. *The Senses of Touch: Haptics, Affects and Technologies*. Oxford: Berg. 2007.

Paterson, Mark, and Martin Dodge. *Touching Space, Placing Touch*. New York: Routledge, 2016.

Pearce, J. M. S. "The Law of Specific Nerve Energies and Sensory Spots." *European Neurology* 54, no. 2 (2005): 115–17. https://doi.org/10.1159/000088647.

Pens, Dan. "Skin Blind." In *Prison Masculinities*, ed. Donald F. Sabo, Terry A. Kupers, and Willie James London, 150–52. Philadelphia: Temple University Press, 2001.

Pettersson, Ingemar. "Mechanical Tasting: Sensory Science and the Flavorization of Food Production." *Senses and Society* 12, no. 3 (February 2017): 301–16. https://doi.org/10.1080/17458927.2017.1376440.

Pink, Sarah. *Doing Sensory Ethnography*. London: Sage, 2009.

Prescott, James W. "Body Pleasure and the Origins of Violence." *Bulletin of the Atomic Scientists* 31, no. 9 (1975): 10–20. https://doi.org/10.1080/00963402.1975.11458292.

Ramachandran, V. S., and David Brang. "Tactile-Emotion Synesthesia." *Neurocase* 14, no. 5 (December 2008): 390–99. https://doi.org/10.1080/13554790802363746.

Reese, Hope. "Americans Are Having Less Sex, but Is That a Problem?" *Greater Good Magazine*, February 18, 2019. https://greatergood.berkeley.edu/article/item/americans_are_having_less_sex_but_is_that_a_problem.

Richards, Rachel. *Hungry for Life: A Memoir Unlocking the Truth Inside an Anorexic Mind*. Self-published, 2016.

Richards, Sabrina. "Pleasant to the Touch." *Scientist*, September 2012.

Ritchie, Ian. "Fusion of the Faculties: A Study of the Language of the Senses in Hausaland." In *The Varieties of Sensory Experience: A Sourcebook in the Anthropology of the Senses*, ed. David Howes, 192–202. Toronto: University of Toronto Press, 1991.

Rigelsford, Jon. "Haptic Human-Computer Interaction: First International Workshop." *Sensor Review* 24, no. 1 (2004). https://doi.org/10.1108/sr.2004.08724aae.003.

Rodgers, Joseph Lee, Craig A. St. John, and Ronnie Coleman. "Did Fertility Go Up After the Oklahoma City Bombing? An Analysis of Births in Metropolitan Counties in Oklahoma, 1990–1999." *Demography* 42, no. 4 (2005): 675–92. https://doi.org/10.1353/dem.2005.0034.

Rothen, Nicolas, and Beat Meier. "Higher Prevalence of Synaesthesia in Art Students." *Perception* 39, no. 5 (2010): 718–20. https://doi.org/10.1068/p6680.

Ruhle, Otto. *Karl Marx: His Life and Works*. New York: Viking, 1943.

Saint-Eve, Anne, Enkelejda Paçi Kora, and Nathalie Martin. "Impact of the Olfactory Quality and Chemical Complexity of the Flavouring Agent on the Texture of Low Fat Stirred Yogurts Assessed by Three Different Sensory Methodologies." *Food Quality and Preference* 15, no. 7–8 (2004): 655–68. https://doi.org/10.1016/j.foodqual.2003.09.002.

Sapolsky, Robert M. *Behave: The Biology of Humans at Our Best and Worst.* London: Vintage, 2018.

Schwenkler, John. "Do Things Look the Way They Feel?" *Analysis* 73, no. 1 (2012): 86–96. https://doi.org/10.1093/analys/ans137.

Shuster, Sam. "The Nature and Consequence of Karl Marx's Skin Disease." *British Journal of Dermatology* 158, no. 1 (January 2008): 1–3. https://doi.org/10.1111/j.1365-2133.2007.08282.x.

Siegel, Reva B. "Introduction: A Short History of Sexual Harassment." In *Directions in Sexual Harassment Law*, ed. Catharine A. MacKinnon and Reva B. Siegel, 1–28. New Haven, Conn.: Yale University Press, October 2003. https://doi.org/10.12987/yale/9780300098006.003.0001.

Simpson, Jeffrey A., and W. Steven Rholes. *Attachment Theory and Close Relationships.* New York: Guilford, 1998.

Smith, Chuck, and Sono Kuwayama. "On Handwork." YouTube video, 4:29. June 26, 2010. https://www.youtube.com/watch?time_continue=2&v=bfoByYLSBY8&feature=emb_logo.

Spanlang, Bernhard, Jean-Marie Normand, David Borland, Konstantina Kilteni, Elias Giannopoulos, Ausiàs Pomés, Mar González-Franco, Daniel Perez-Marcos, Jorge Arroyo-Palacios, Xavi Navarro Muncunill, and Mel Slater. "How to Build an Embodiment Lab: Achieving Body Representation Illusions in Virtual Reality." *Frontiers in Robotics and AI* 1 (November 27, 2014). https://doi.org/10.3389/frobt.2014.00009.

Steiner-Adair, Catherine. *The Big Disconnect: Protecting Childhood and Family Relationships in the Digital Age.* New York: Harper, 2014.

Sullivan, Andrew. "I Used to Be a Human Being." *New York Magazine*, September 19, 2016, https://nymag.com/intelligencer/2016/09/andrew-sullivan-my-distraction-sickness-and-yours.html.

Twenge, Jean M., Ryne A. Sherman, and Brooke E. Wells. "Sexual Inactivity During Young Adulthood Is More Common Among U.S. Millennials and IGen: Age, Period, and Cohort Effects on Having No Sexual Partners After Age 18." *Archives of Sexual Behavior* 46, no. 2 (January 2016): 433–40. https://doi.org/10.1007/s10508-016-0798-z.

Udry, J. Richard. "The Effect of the Great Blackout of 1965 on Births in New York City." *Demography* 7, no. 3 (1970): 325. https://doi.org/10.2307/2060151.

Van der Kolk, Bessel, *The Body Keeps the Score: Brain, Mind and Body in the Healing of Trauma.* New York: Penguin, 2015.

Vignemont, Frederique De. "Pain and Bodily Care: Whose Body Matters?" *Australasian Journal of Philosophy* 93, no. 3 (2014): 542–60. https://doi.org/10.1080/00048402.2014.991745.

Wade, Nicholas J. *A Natural History of Vision*. Cambridge, Mass.: MIT Press, 1999.

Willis, Ellen. "Toward a Feminist Sexual Revolution." *Social Text*, no. 6 (1982): 3. https://doi.org/10.2307/466614.

Wilson, Frank R. *The Hand: How Its Use Shapes the Brain, Language, and Human Culture*. New York: Vintage, 1999.

Wolf, Maryanne. *Proust and the Squid: The Story and Science of the Reading Brain*. New York: Harper, 2007.

Wong, Michael, Vishi Gnanakumaran, and Daniel Goldreich. "Tactile Spatial Acuity Enhancement in Blindness: Evidence for Experience-Dependent Mechanisms." *Journal of Neuroscience* 31, no. 19 (November 2011): 7028–37. https://doi.org/10.1523/jneurosci.6461-10.2011.

Wujastyk, Dominik. *The Roots of Ayurveda: Selections from Sanskrit Medical Writings*. London: Penguin, 1998.

Zarra, Ernest. *Teacher-Student Relationships: Crossing into the Emotional, Physical, and Sexual Realms*. Plymouth, U.K.: R & L Education, 2013.

Zimmer, Carl. *Smithsonian Intimate Guide to Human Origins*. New York: Harper Perennial, 2007.

Zur, Ofer. "Touch in Therapy." In *Boundaries in Psychotherapy: Ethical and Clinical Explorations*, 167–85. Washington, D.C.: American Psychological Association, 2007. https://doi.org/10.1037/11563-010.

Index

A delta fibers, 36
active touch, 27. *See also* haptic perception
Age of Excarnation, 92
AIDS, 137
Ainu, 213
acupuncture, 156
amputation, 190
analog, 181–82
anorexia nervosa, 87–88
Apple, 172
Aristotle, xi, 57–58
attachment theory, 98–100
autism, 110
Ayurveda, 154, 156

baby boom, 136. *See also* blackout baby
Berkeley, George, 81
Bible, the, 213
Biden, Joe, 128–29
Biotac, 190–93, 197. *See also* SynTouch
blackout baby, 136. *See also* baby boom
Brave New World (Huxley), 187
breastfeeding, 123
Broca's area, 201
Brooks, Fred, 172

capitalism, 4, 139–40, 165. *See also* consumerism
Carlson, Mary, 108
Casino de Montreal, 22
Ceaușescu, Nicolae, 107
C fibers, 36
childrearing, 20, 98–100, 103–10, 123, 214. *See also* parenting
consumerism, 4, 139–40, 165. *See also* capitalism
coronavirus, 208–11. *See also* Covid-19

C-tactile afferents, 43
child development, 39–40, 201, 206–7
cortisol, 109
Covid-19, 208–11. *See also* coronavirus
crossmodal associations, 147–49
cuddler, 119–22, 137–41. *See also* professional cuddler
cultural differences, 12–14, 128, 160–62

Darwin, Charles, 99
dating, 93, 127–28, 133
digital era, 21–22, 75, 90–94, 142; digital detox, 91–92
disorder, 126–27, 40, 44–45
domestic violence, 124
dopamine, 109
driverless cars, 204
Dunbar, Robin, 201

Eisenberger, Naomi, 64–66
emotion, 64–69
Enlightenment, xii, 17–19
e-reader, 181–82
extroversion, 98
evolution, 199–202

feminism, 123–26
Field, Tiffany, 108–9
fight-or-flight response, 108, 109
flotation, 162
flow, 207
fur, 163–64

gender, 122–26, 129; femininity, 16; masculinity, 131–35
Goertz, Ray, 171–72
Grandin, Temple, 111
Gromala, Diane, 73–78

Hagbarth, Karl-Erik, 42–43
Handmaid's Tale, The (Atwood), 186
haptic: engineering, 167–69; perception, 27; World Haptics Conference, 170–76. *See also* active touch
Harlow, Harry, 105–7, 123
Hausa, 9
Hayward, Vincent, 177
hereditary sensory neuropathy, 40
Highsmith, Patricia, 186
Hill, Anita, 125
Hinduism, 214
histamine receptor, 35
homunculus, 37
Howes, David, 4–10
human resources, 126
Huxley, Aldous, 187

illusion, 83–85, 177–81; mirror box, 88–89; rubber hand, 85–87
immune system, 27
incel, 134–35
industrial Age, xii, 19–21, 151
insula, 37–38
introversion, 98

Kandinsky, Wassily, 67
Kawabata evaluation system, 143–44

Kearney, Richard, 92–95
King Lear, 85

lanceolate ending, 34
law of specific nerve energies, 79. *See also* Müller, Johannes
language, 201–2
Leibniz, Gottfried, 82
Locke, John, 80–81
loneliness, 134

Maker Movement, 165
Malloy, Julie, 40–41
marketing, 145, 165
Marx, Karl, 1–4
marriage, 132
massage, 96–103, 113–16, 118, 154–55, 219
Massim, 8–10
medicine, 17, 94, 154–59; alternative, 156–59
mechanoreceptor, 34–36; lanceolate ending, 34; Meissner's corpuscle, 34; Merkel cell, 34; Pacinian corpuscle, 34; Ruffini ending, 34. *See also* sensors
Meissner's corpuscle, 34
#metoo movement, xiii, 126
Merkel cell, 34
metaphor, 67–69, 84–85
microneurography, 42
Middle Ages ,15–17
Middle Sepik, 9–10
mirror box illusion, 88–89
Milton, John, 84
Molyneux's question, 80–83
Morrison, India, 45
Moseley, Lorimer, 88
movement, 27, 38, 40, 192
Müller, Johannes, 79. *See also* law of specific nerve energies
Munthe, Axel, 156

Nabokov, Vladimir, 67
NBA, 116–17
nerves, 27, 36–37, 42–45; A delta fibers, 36; C fibers 36; C-tactile afferents, 43
Nestlé, 146–50
New Age, 164
nociceptor, 35
nonverbal communication, 12–15, 116–17

Obama, Michelle, 128
objectivity, 151, 153
Oklahoma City bombing, 136
Olausson, Håkan, 42–43
oxytocin, 109

Pacinian corpuscle, 34
pain, 27, 36, 157–58, 180, 218; chronic, 73–78, 88–89
Papua New Guinea, 7–10
parasympathetic nervous system, 109
parenting, 20, 98–100, 103–10, 123, 214. *See also* childrearing
posterior parietal cortex, 38
Plato, x–xi
pornography, 94
priming, 64

prison, 20, 130
professional cuddler, 119–22, 137–41. See also cuddler
proprioception, 27, 38, 40
prosthesis, 190–98
Protestantism, 155

Queen Elizabeth, 128

race, 19
rape, 124
Ramachandran, V. S., 50, 67; mirror box illusion, 88–89
rehabilitation, 30–31, 112
religion, 13, 17, 18, 155, 160; Hinduism, 214; Protestantism, 155. *See also* spirituality
robot, 204–6
rubber hand illusion, 86
Ruffini ending, 34

sensor, 34–36; lanceolate ending, 34; Meissner's corpuscle, 34; Merkel cell, 34; Pacinian corpuscle, 34; Ruffini ending, 34. *See also* mechanoreceptor
sensory ethnography, 5–10, 211–14
sensory evaluation, 144–53
sensory hierarchy, x, 19
serotonin, 109
sex, 47, 124, 127, 132–33, 140; robots, 141; sex work, 140
sexual harassment, 124–25
Shakespeare, 85
skin: hunger, 110; physiology, 33–37; skin-to-skin, 110
somatic awareness practice, 111–13
spa, 142, 165
spirituality, 13, 17, 18, 155, 160; Hinduism, 214; Protestantism, 155. *See also* religion
stretch, 34
synesthesia, 49–52, 59
SynTouch, 190–93, 197. *See also* Biotac

Taung Child, 199
temperature, 27, 35, 65
textile, 143–45
thalamus, 37
thermoreceptor, 35
touch avoidance, 98
Tyler, Dustin, 193–98

Vallbo, Åke, 42
vibration, 34, 36
virtual reality, 73–78, 89–92, 175
vision, cooperation between touch and, 80–85
Volkswagen, 167–70, 172

Waterman, Ian, 25–33, 41–42, 45–48
Watson, John B., 104–5
Williams, Alicia Elba, 49–57, 69–72
work, culture of, 15–17, 19

yoga, 112, 159, 165

ABOUT THE AUTHOR

Sushma Subramanian is an associate professor of journalism at the University of Mary Washington. Her writing has appeared in *Slate*, *The Atlantic*, *New York*, and *Discover*, among others.